I0489306

CASE STUDY ANALYSIS OF THE U.S. AND EU PROPOSALS

Task Final Report

June 2001

Prepared for:

US Department of Energy
National Energy Technology Laboratory
626 Cochrans Mill Road
Pittsburgh, PA 15236

Prepared by:

Science Applications International Corporation
8301 Greensboro Drive
McLean, VA 22102

TABLE OF CONTENTS

LIST OF TABLES

LIST OF FIGURES

1. EXECUTIVE SUMMARY

Introduction

The issue of greenhouse gas emissions has been at the forefront of environmental concerns for the past decade. A number of treaties, agreements, and voluntary programs have been proposed to reduce emissions – some of which have been the subject of intense debate and disagreement. Most notable among these proposals has been the Kyoto Protocol. Signed in 1997 by the United States and other industrialized countries, the Kyoto Protocol is a major international treaty imposing binding emission reduction targets on the developed world. However, the U.S. Senate never ratified Kyoto, and the Administration recently announced its intention of dropping out of the international negotiations surrounding the Protocol. Nonetheless, the general scientific consensus, that global warming is a real, significant issue, is not in dispute. The Administration is calling into question only the appropriate response to this issue, while explicitly recognizing the need for *some* response.

Regardless of whether this response takes the form of a domestic voluntary program, an international treaty, or something in between these two extremes, it is likely that it will incorporate "market mechanisms" in some form or other. Most of the various emission reduction responses that have been proposed over the past few years include such mechanisms. The development and implementation of these mechanisms, designed to facilitate low-cost solutions to environmental problems, is part of a broader trend away from the command-and-control regulations of the past, and towards increased flexibility in meeting regulatory requirements. This new market-based approach has worked it way into greenhouse gas emission reduction programs and proposals, using the guidelines provided by the United Nations Framework Convention on Climate Change (UNFCCC), and developed into a new concept: credits for emission reduction projects undertaken beyond a country's borders.

Perhaps the greatest challenge for this new concept is the development of a protocol, or set of protocols, for estimating the emission reductions associated with projects. There is considerable concern among various groups surrounding the accuracy of the emission reduction estimates upon which credits would be awarded. In addition, others, particularly any potential project developer, want protocols that can be implemented within reasonable costs. Nonetheless, all parties generally recognize the need for accuracy of credits and agree on the need for a standard approach or set of procedures for estimating project-level emission reductions. A number of such approaches have been proposed and the purpose of this report is to evaluate some of the key proposals. Specifically, the report presents a series of hypothetical case study analyses designed to test each proposed approach in the context of potential real world projects. The case studies have been selected to cover a broad range of sectors and project types. The goal is to identify the strengths and weaknesses of each approach, and based on the case study analyses, recommendations for improving and refining the different approaches are developed.

Four different approaches are evaluated in this report:

- The approach officially proposed by the U.S. at the recent (COP-6) negotiations surrounding the Kyoto Protocol

- The European Union's "Positive Technology List"

- The U.S. National Energy Technology Laboratory's (NETL) technology matrix concept (the "full" technology matrix)

- A hybrid approach combining elements of the technology matrix with the official U.S. approach (the "hybrid" technology matrix)

The Case Studies and the Methodological Approach

Each case study project is evaluated using each of the above four approaches. The results for each approach are analyzed, compared and contrasted; these critical analyses in turn reveal the strengths and weaknesses of the different approaches in the context of a variety of different project types.

The Four Approaches

The Official U.S. Approach. Although initially proposed during the negotiations on the Kyoto Protocol, the "official" U.S proposal remains relevant (despite the uncertain future of the Protocol) as a potential starting point for any future approach crafted to meet the needs of either a voluntary domestic program, or an international mandatory agreement. It suggests a two-step approach to dealing with additionality and baseline development. In the first step, a project's eligibility for credits is determined through a comparison of the project's emissions with a standard benchmark representing a level of emissions performance that is significantly better than the average for recent, comparable projects. In the second step, the credits to be awarded to qualifying projects would be computed by subtracting the project's emissions from a second benchmark, representing the average emissions of recent, comparable projects.

The EU Positive List. The EU has proposed that only projects based on a "positive list" of safe, environmentally sound, clean technology projects should be able to obtain credits.[1] The proposed positive list is presented in Table ES1.

---

[1] European Commission, "Outcome of Climate Change Negotiations in Lyon, France, 4-15 September, 2000 (Press Release)," September 1, 2000, http://europa.eu.int/comm/environment/press/bio00172.htm

Table ES-1. The EU's Positive List of Technologies

Main Technology Categories	Individual Technologies
Renewables	Solar
	Wind
	Sustainable Biomass
	Geothermal heat and power
	Small-scale hydropower
	Wave and tidal power
	Ambient heat
	Biogas
Energy Efficiency	Advanced technologies for combined heat and power installations and gas fired power plants
	Significant improvements in existing energy production
	Advanced technologies for, and/or significant improvements in industrial processes, buildings, energy transmission, transportation and distribution
	More efficient and less polluting modes of mass and public transport (passenger and goods) and improvement or substitution of existing vehicles
Demand Side Management	Improvements in residential, commercial, transport and industrial energy consumption.

The Full Technology Matrix. The technology matrix approach, modified and developed by the U.S. Department of Energy's National Energy Technology Laboratory (NETL) consists of a selected list of greenhouse gas abating technologies that correspond with the sustainable development goals of a host country.[2] Additionality and baseline determination, under this approach, take place in two stages. First, a technology is subjected to an additionality test to determine whether it should be included in the matrix. This test would be based on factors such as the commercial viability and market penetration of the candidate technology. The test will be designed to ensure that only advanced, non-commercial technologies qualify for inclusion in the matrix

Second, a stipulated benchmark will be developed for each approved technology based on the emissions performance of a selected group of counterfactual technologies. To qualify for credits, project developers would simply demonstrate that the proposed project technology is included in the matrix. The stipulated benchmark from the matrix would then be used to calculate the project's emission reductions.

The Hybrid Technology Matrix. The hybrid technology matrix approach is based on a combination of the full technology matrix's additionality test and the second step of the official U.S. proposal for baseline development.

---

[2] As sustainable development criteria are likely to vary among countries most examples of the modified technology matrix are anticipated to be country-specific.

The Case Studies

Each of the above-described approaches is applied to forty case studies. In developing the case studies, the objective has been to cover a variety of plausible projects in an attempt to test the four approaches against a full spectrum of situations likely to arise under a future carbon mitigation regime.

Table ES2 lists the case studies. In this table, the case studies are organized by sector with eleven case studies developed for the electricity sector, thirteen for the industrial sector, nine for the transportation sector, three for the residential sector, two for the commercial sector and two for the forestry sector. Because the electricity sector has received more attention in the development of the technology matrix and the U.S. approach, we have shifted the emphasis somewhat towards other sectors. The goal is to test the methodologies in applications, which are plausible under future carbon mitigation regimes. All of the case studies identified in Table ES2 are fictitious. In addition, most (although not all) of the "data" utilized in the case studies are fictitious. The use of hypothetical projects, with fictitious data, significantly reduced the amount of time required to develop each case study. This in turn enabled the development of a large number and variety of case studies—a key objective of the analysis, given the desire to test the methodologies under the full spectrum of plausible scenarios. Had an attempt been made to obtain actual data for the case studies, the data collection effort would have drastically reduced the amount of time available for case study development and analysis. Furthermore, in many cases it would likely have proved impossible to obtain the required data.

Table ES-2. The Case Studies

Sector	Project ID #	Country	Project Title
Electricity	ES1	India	IGCC Power Plant
	ES2	India	Heat Rate Improvement
	ES3	India	Fuel Switching
	ES4	India	Natural Gas Combine Cycle
	ES5	India	Gas Turbine Plant
	ES6	India	Wind Power
	ES7	Kazakhstan	IGCC in Kazakhstan
	ES8	Tajikistan	Hydropower
	ES9	India	Distributed Generation: Fuel Cells
	ES10	China	Transmission Capacity Expansion
	ES11	India	Carbon Sequestration for IGCC Plant
Industrial	IS1	Azerbaijan	Installation of District Heating System
	IS2	Kazakhstan	Cogeneration at Food Processing Plant
	IS3	Argentina	Variable Frequency Drives
	IS4	Brazil	Retrofit of Energy Efficient Motors
	IS5	China	Coke Oven Underfiring Rate Improvement
	IS6	Tajikistan	PFC Reductions at Aluminum Plant
	IS7	China	Coal Ash Utilization
	IS8	Chile	Building Insulation Improvement
	IS9	Jordan	Highly Efficient Fertilizer Complex
	IS10	China	Industrial Boiler Shutdown
	IS11	South Africa	Coal Mine Methane Recovery

	IS12	Argentina	Landfill Gas Flaring
	IS13	Kazakhstan	Recovery of Associated Natural Gas
Transportation	TS1	India	Dedicated CNG Taxis
	TS2	India	New Gasoline-Fueled Taxis
	TS3	China	Aluminum Rail Cars for Efficient Coal Transport
	TS4	S. Africa	Clean Diesel in Transit Buses
	TS5	Mexico	Electric Vehicles in Mexico City
	TS6	Thailand	Smart Toll System
	TS7	Ukraine	46 New Conventional Diesel Buses
	TS8	India	New Two-Wheelers
	TS9	Brazil	Improving Road Infrastructure
Land Use	LU1	Mexico	Forest Protection and Management
	LU2	Russian Federation	Afforestation of Marginal Agricultural Land in Russia
Residential	RS1	South Africa	Construction of Energy-Efficient Homes in South Africa
	RS2	Mexico	Sale of High-Efficiency Light Bulbs for Homes
	RS3	Russian Federation	Energy Efficiency of Seven Apartment Buildings
Commercial	CS1	Philippines	Energy Efficiency and Conservation Measures in Commercial Buildings
	CS2	Indonesia	Motor Replacement Project in Commercial Office Buildings in Jakarta

Each of the case studies listed in Table ES2 is analyzed using each of the four emission credit estimation approaches. Under each approach, a determination is made to whether or not the project should qualify for credits or be rejected as a free rider. Then, if the project qualifies under a given approach, the credits that would be awarded to the project under the approach are estimated. Finally, the results of this analysis are subjected to a critique, in order to identify the strengths and weaknesses of each approach vis a vis the particular project.

Summary and Lessons Learned

Based on the detailed case studies, a number of main conclusions can be drawn, as follows:

- All of the project evaluation approaches demonstrate the capacity to misclassify free rider projects as additional (and vice versa). Often, these qualification errors differ among the approaches, making generalizations regarding project type difficult. However, in general, whereas the U.S. approach typically fails by qualifying free rider projects as additional, the EU and technology matrix approaches tend to fail by misclassifying truly additional projects as free riders. Also, the technology matrix approaches appear to result in the fewest qualification errors, although it is cautioned that this conclusion is based on an examination of hypothetical case studies that may not be representative of actual, future projects.

- Each of the four approaches encountered situations in which they simply could not be applied. The EU positive list encountered the most difficulties: seven case studies that simply could not be analyzed using the EU approach. This problem was found to arise from the vague, imprecise language used to define technologies within the positive list. The U.S. and technology matrix approaches also simply broke down in a number of cases. These failures manifest the need to include a backup methodology as an explicit default for *any* standardized, multi-project approach ultimately adopted. This back-up approach should be an ad hoc, non-standardized procedure that can be tailored to the characteristics of any particular project.

- Both the U.S. approach and the technology matrix approaches require the existence of facilities comparable to the project being assessed. The emissions data for these comparable facilities are used to benchmark the project. However, for some types of projects, and some countries, comparable facilities are likely to prove nonexistent. For example, this problem arose frequently for the countries of the FSU. Due to the long-term economic decline these countries have experienced, there is a dearth of recently built power plants and other facilities against which new projects can be compared. In addition to certain countries, the problem of nonexistent comparable facilities appears to plague certain sectors more often then others. For example, comparable facilities proved difficult or impossible to identify for a number of industrial sector case studies, due to the heterogeneous nature of projects in the industrial sector.

- The data required to perform the project analyses is, in many cases, likely to prove unavailable. Data availability will be a particular problem for the U.S. approach, because this approach requires the development of a percentile distribution of emission rate data for comparable facilities. The data requirements of the technology matrix approaches are less stringent, although even for these methods data availability is likely to prove a major problem for many developing country projects.

- The EU positive list is clearly less developed and well-defined than the other three approaches tested. A number of major problems arose from the application of the positive list to the case studies. First, the positive list lacks sufficient clarity in its definition of qualifying technologies and processes. Second, some projects were found to fit under more than one category on the list. The fact that a single project could potentially fall into two separate categories in the positive list, resulting in potentially conflicting qualification determinations, is clearly a fundamental internal inconsistency. Third, the EU approach fails to provide a procedure for quantifying the credits to be awarded to qualifying projects. Finally, the positive list focuses exclusively on energy-related projects, thereby automatically disqualifying whole classes of important projects. For example, the positive list automatically disqualifies projects aimed at reducing HFCs, PFCs and sulfur hexafluoride, despite the fact that these are very potent greenhouse gases.

- None of the four approaches provide adequate guidance for handling land use and forestry sector projects.

A number of recommendations were developed for addressing the above-noted problems and improving each of the four methods. In the case of the EU's positive list, improvements can be realized by (1) clarifying the definitions of qualifying technologies as well as quantifying qualification criteria; (2) developing a methodology for quantifying the number of credits to be awarded to qualifying projects; and (3) expanding the positive list to include non-energy, non-carbon related project opportunities. In the case of the U.S. approach, the distinction between new facility and retrofit projects needs to be clarified, and a backup additionality test that accounts for projects that fall outside the "efficiency and/or emission rate" box should be established. Finally, the two technology matrix approaches could be improved by strengthening the market penetration test to provide effective evaluation of first-of-its-kind projects, and explicitly addressing the treatment of multi-component projects that utilize advanced technologies for only some of the components.

Conclusions

The case study analyses indicate that, of the four approaches tested, the technology matrix provides the most stringent additionality test. Furthermore, the technology matrix offers several other advantages. First, it explicitly incorporates an alternative, project-specific backup methodology to be used in situations where the matrix does not apply or is unable to provide an accurate emission reduction estimate. Furthermore, the technology matrix is less data-intensive than the official U.S. approach. Finally, the technology matrix is technology neutral in the sense that it focuses on the additionality of the activities examined rather than relying on political processes to determine an emissions threshold or an acceptable technology.

2. INTRODUCTION

Background

The issue of greenhouse gas emissions has been at the forefront of environmental concerns for the past decade. A number of treaties, agreements, and voluntary programs have been proposed to reduce emissions. Some of these proposals have been the subject of intense debate and disagreement. Most notable among these proposals has been the Kyoto Protocol. Signed in 1997 by the United States and other industrialized countries, the Kyoto Protocol is a major international treaty imposing binding emission reduction targets on the developed world. However, the U.S. Senate never ratified Kyoto, and the Administration recently announced its intention of dropping out of the international negotiations surrounding the Protocol.

Nonetheless, the general scientific consensus, that global warming is a real, significant issue, is not in dispute. The Administration is calling into question only the appropriate response to this issue, while explicitly recognizing the need for *some* response.

Furthermore, regardless of whether this response takes the form of a domestic voluntary program, an international treaty, or something in between these two extremes, it is likely that it will incorporate "market mechanisms" in some form or other. Most of the various emission reduction responses that have been proposed over the past few years include such mechanisms. The development and implementation of these mechanisms, designed to facilitate low-cost solutions to environmental problems, is part of a broader trend away from the command-and-control regulations of the past, and towards increased flexibility in meeting regulatory requirements.

The Development of Market Mechanisms: A Historical Overview

An early example of this trend away from the old command-and-control approach was the U.S. sulfur dioxide emission allowance trading system. The SO_2 emission allowance market has been operating successfully in the U.S. for a number of years, and is widely credited as having reduced the costs of compliance with SO_2 emission regulations. Under this system, an overall cap on SO_2 emissions is set, but the market is empowered to allocate the emission reductions required under this cap to individual electricity suppliers in the most cost-effective manner possible. Emission allowances are issued in an amount equal to the cap; these allowances can be traded. Suppliers that can meet and exceed their SO_2 reduction requirements at costs below the market clearing price of emission allowances are afforded the opportunity to sell their excess allowances, while suppliers facing SO_2 reduction costs above the market clearing price are able to meet their requirements by purchasing allowances. Notably, the SO_2 allowance trading system has proved to be a political, as well as an economic and environmental success—it has gained widespread support among environmentalists as well as industry.

The success of the U.S. SO$_2$ allowance trading system led to the incorporation of similar, international emission allowance trading systems under the UNFCCC and other greenhouse gas reduction programs. Furthermore, the new market-based approach exemplified by emission allowance trading was extended by the development of a new concept: credits for emission reduction activities undertaken beyond a country's borders. The exemplar of this new market mechanism is the Clean Development Mechanism (or CDM). A CDM activity is defined as a project between a developed country and a developing country, in which the developing country plays host to the project, while the developed country provides financial and/or technical support to the project in exchange for emission reduction credits. These credits can be applied to the developed country's emission reduction obligations, or alternatively they can be traded on the international market.

In addition to the CDM, a similar market mechanism, known as Joint Implementation (or JI) was developed. JI projects are the same as CDM projects, except that the former projects are implemented by developed country partners, rather than developed/developing country partners. The CDM and JI are intended to provide "win-win" situations for both the host country and its partner. The host country receives financial and/or technical backing for projects designed to meet its energy (or other) needs, while the other partner receives emission reduction credits in exchange for its financial/technical aid. It is generally believed that these developing country opportunities are substantial, and, furthermore, that many of them can be implemented at lower cost than emission reduction activities undertaken within the U.S. For example, simply refurbishing and modernizing old, worn-out power plants in the developing world, and bringing them up to U.S. efficiency levels, could have a major impact on global emissions at relatively low cost. Hence, the CDM was viewed, within the U.S., as a means of helping us to meet our emission reduction obligations while minimizing our costs of compliance.

Beyond this economic reason, there was a second, political reason for U.S. interest in the CDM. Some parties involved in the international climate change negotiations maintained that developing countries should not be required to reduce emissions. The exclusion of developing countries from emission reduction obligations proved to be a major obstacle to ratification of the Kyoto Protocol by the U.S. Senate. Developing, rapidly growing countries such as India and China are expected to account for most of the growth in global emissions over the coming decades. Many parties argued that any international agreement that fails to impose emission reduction obligations on developing countries is fundamentally flawed. Others countered this argument by pointing out that the CDM ensures developing country participation in meeting the agreed-upon emission reduction goals. The basic concept of the CDM does not limit itself to the context of the Kyoto Protocol. Its practical and market-based approach is likely to be adopted under other policy alternatives as discussed in subsequent sections.

The Future of Market Mechanisms

As already noted, the United States has signaled its intention of dropping out of the negotiations surrounding implementation of the Kyoto Protocol. At the same time, the Administration has made clear its intention of pursuing other policy alternatives.

It is unclear, at this point, what form these alternatives might take. The possibilities range from relatively simple, voluntary domestic programs, to complex international treaties imposing mandatory emissions targets. At this point, it appears more likely that the U.S. response to climate change will tend towards the former. However, this could well change as the politics of the climate change issue continue to evolve.

But whether the United States ultimately adopts a voluntary domestic approach, a mandatory international approach, or something in between, it seems highly likely that the approach will prominently feature market mechanisms. In particular, the CDM concept of credits for emission reduction activities undertaken beyond U.S. borders is likely to be incorporated in any U.S.-implemented approach. Both the economic and political advantages of this concept will continue to hold, within or beyond the context of the Kyoto Protocol. Many low-cost emission reduction opportunities no doubt exist beyond U.S. borders; the American business community will lobby to receive credit for exploiting these opportunities, and this lobbying effort will more than likely prove successful. Any program that encourages developing countries to reduce emissions—and perhaps most especially a voluntary, primarily domestic U.S. program—will be well-received within the United States.

And furthermore, the precedent for market-based approaches, and in particular CDM-type programs—is already well established. The U.S. government already administers a number of voluntary emission reduction programs, including, e.g., the Climate Challenge program, the Climate Wise program, and the Voluntary Reporting of Greenhouse Gases program. Under the Climate Challenge program, electricity suppliers enter into voluntary agreements with the U.S. Department of Energy to reduce their emissions; the magnitude of the agreed-upon reductions varies from agreement to agreement. The Climate Wise program is similar to Climate Challenge, except that it applies to other sectors beyond the electricity sector. Under both programs, volunteers are allowed to apply the emission reductions they achieve beyond U.S. borders to their agreed-upon emission reduction targets.

The Voluntary Reporting of Greenhouse Gases (VRGG) program, administered by the Energy Information Administration, allows any U.S. entity to report their emission reduction achievements on a voluntary basis. As in the case of the Climate Challenge and Climate Wise programs, credits are not awarded for emission reductions reported to the VRGG program. However, many of the reporters who participate in the program are motivated by a desire to register their *claims* to possible future credits. In effect, the VRGG serves as a public registry of claims to future emission reduction credits. VRGG reporters are explicitly allowed to report their claims to foreign as well domestic emission reductions. Furthermore, the VRGG allows reporters to report purchases and sales of emission reductions.

A number of VRGG reporters are reporting such trades. Indeed, despite the absence of any mandatory emission reduction program in either the U.S. or abroad, a nascent, international emission reduction market now exists, with buyers, sellers, and brokers. The commodity being traded on this market is not entirely clear. Emission allowances, of the type traded on the U.S. SO_2 market, are *not* being bought or sold on the nascent greenhouse gas market. Neither the U.S. government nor any other government has issued such allowances. Rather than *allowances* to *emit* emissions, the commodity currently being traded is related to actual *emission reductions* achieved either through a specific project, or through a group of projects or activities. Participants in the market may tend to view the actual traded commodity as *credits* for these project-based emission reductions. However, since such "credits" have not as yet received any official sanction, they are perhaps better viewed as *claims* to credits.

But granted that the commodity being traded is claims to future credits rather than credits themselves, and that the market for these claims remains at this point quite small and undeveloped, it is nonetheless remarkable that the market exists. As yet there are no mandatory emission reduction programs, in the either U.S. or abroad, which can sanction the validity of the claims, let alone provide the context within which the claims would have value. The development of a nascent international market in project-level emission reduction claims, despite these adverse conditions, attests to the early acceptance and support of the CDM concept within the business community. Similarly, the U.S. government's Climate Challenge, Climate Wise, and VRGG programs set important precedents for the concept of credits for foreign emission reductions. Participants in these three programs include most of the major U.S. electric utilities, as well as many other major American corporations (including General Motors, IBM, Dow, and Johnson & Johnson, to name a few). The individual claims of these program participants, and of emission reduction traders, are likely to be scrutinized in the context of any future mandatory program. Some of these claims might even be rejected as unworthy of credit, for a variety of reasons. However, at this point it seems unlikely that the *concept* of credits for project-level emission reductions—including foreign reductions—will be rejected. On the contrary, the demonstrated support for this concept within both the government and corporate America makes it difficult to imagine any future emission reduction program—voluntary or mandatory, domestic or international—that does not include some form of credits for foreign emission reduction projects. The political momentum clearly favors the market-based approach over the old command-and-control approach, and the CDM concept has quickly found favor as a key component of any market-based approach.

Objectives and Report Organization

In short, regardless of the future of the Kyoto Protocol, the CDM concept is likely to live on in future emission reduction programs. Perhaps the CDM concept will be given a new name and an altered form, although major alterations to the basic idea—credits for foreign emission reduction projects—seem unlikely. However, this is not to say that the CDM concept is not without its challenges. Perhaps foremost among these challenges is the development of a protocol, or set of protocols, for estimating the emission reductions associated with projects. Although U.S.-based environmental groups generally support the basic concept of market mechanisms and the CDM, there is considerable concern within the environmental community surrounding the accuracy of the emission reduction estimates upon which credits would be awarded. Balanced against the environmentalists' demands for protocols that will ensure the integrity of the credits, the business community wants protocols that can be implemented within reasonable costs. Nonetheless, corporations and participants in the nascent emission reduction market generally recognize the need for accuracy, in order to ensure that claims to credits will withstand scrutiny and ultimately be recognized as valid.

In any event, both environmental groups and corporations agree on the need for a standard approach or set of procedures for estimating project-level emission reductions. A number of such approaches have been proposed. The purpose of this report is to evaluate some of the key proposals. Specifically, the paper presents a series of hypothetical case study analyses. The case studies are designed to test each proposed approach in the context of potential real world projects. The case studies have been selected to cover a broad range of sectors and project types. The goal is to identify the strengths and weaknesses of each approach. Based on the case study analyses, recommendations for improving and refining the different approaches are developed.

Four different approaches are evaluated in this report:

- The approach officially proposed by the U.S. at the recent (COP-6) negotiations surrounding the Kyoto Protocol

- The European Union's "Positive Technology List" approach

- The U.S. National Energy Technology Laboratory's (NETL) technology matrix concept

- A hybrid approach combining elements of the technology matrix with the official U.S. approach.

Although the first two approaches listed above were initially proposed during the negotiations on the Kyoto Protocol, they remain relevant despite the uncertain future of the Protocol. In particular, the official U.S. approach could well serve as the starting point for any future approach crafted to meet the needs of either a voluntary domestic program, or an international mandatory agreement.

The organization of this report is as follows. Chapter 2 provides a brief overview of the four proposed approaches, a summary of the case studies, and a description of the basic methodology used to develop and analyze the case studies. Chapter 3 presents a summary of the key issues and "lessons learned" from the individual case studies. Included in this chapter are our recommendations for improving the different proposals, as well as a discussion of NETL's technology matrix approach as the technical solution to many of the problems posed by the requirements of project evaluation. Appendix A presents the individual case studies, and Appendix B presents more detailed descriptions of the proposed approaches.

3. THE CASE STUDIES AND THE METHODOLOGICAL APPROACH

The Four Methods

Four different approaches to project evaluation are analyzed in this paper:

- The "official" U.S. approach (proposed by the U.S. delegation at COP-6)

- The EU's positive technology list

- The "full" technology matrix

- The "hybrid" technology matrix.

Each case study project included in Appendix A is evaluated using each of the above four approaches. The results for each approach are analyzed, compared and contrasted; these critical analyses in turn reveal the strengths and weaknesses of the different approaches in the context of the variety of project types.

The official U.S. approach, the EU's positive technology list, and the technology matrix are described, in detail, in Appendix B. In the following subsections, a brief overview of each of the four approaches is provided.

The Official U.S. Proposal

Although the official U.S. approach was initially proposed during the negotiations on the Kyoto Protocol, it remains relevant (despite the uncertain future of the Protocol) as a potential starting point for any future approach crafted to meet the needs of either a voluntary domestic program, or an international mandatory agreement.

The official U.S. proposal suggests a two-step approach to dealing with additionality and baseline development. In the first step, a project's eligibility for credits is determined through an additionality test. The second step forms the basis for quantifying the amount of credits to be awarded.

The U.S. approach to additionality is based on a concept of superior performance. To qualify as additional a project activity must achieve a level of performance that is significantly better than average compared with recently undertaken activities or facilities.[3] In other words, the project activity must meet an eligibility threshold that is

---

[3] UNFCCC, Mechanisms Pursuant to Articles 6, 12 and 17 of the Kyoto Protocol, Principles, modalities, rules and guidelines for the mechanisms under Articles 6, 12 and 17 of the Kyoto Protocol Paper No. 4 Unites States of America, September 9, 2000, p. 21. Document # FCCC/SB/2000/MISC.4/Add.1.

Figure 1. Percentile Test

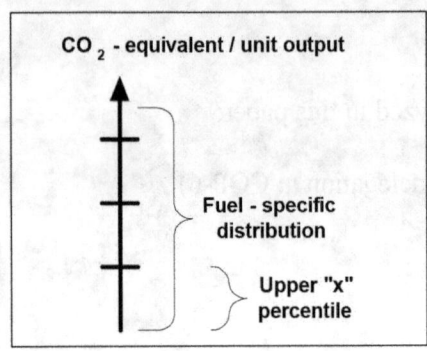

Source: U.S. Environmental Protection Agency Presentation, SB-13 Lyon, France September. 2000

significantly better than average compared to a reference scenario. A reference scenario would consist of a set of recent and comparable activities or facilities in a relevant geographic area, normally a host country. As Figure 1 demonstrates, a certain percentile of the reference scenario will determine the eligibility threshold for each sector or activity type. It is this percentile that the project activity must outperform in order to meet the eligibility threshold and qualify as additional. The requirement that the project meet the percentile threshold defines the criteria that the project achieve a level of performance that is "significantly better than average."

Once the eligibility threshold is met, credits would be granted on the basis of what would have happened in the absence of the project activity. The goal is to create general standards for specific project categories and regions that are both realistic and practical. The reference scenario developed as part of the additionality test would, in many cases, satisfy this criterion but for fossil projects, project developers would be allowed to choose from three categories of baselines 1) a fuel-specific weighted average; 2) a fossil weighted average which would include coal, oil, and natural gas plants; and 3) a sector weighted average, which would include fossil, hydropower, nuclear and renewable plants. Under the U.S. proposal, the number of credits awarded would be determined by subtracting the project activity emissions from the average emissions of the reference scenario or, in the case of fossil projects, one of the three baseline categories.

The U.S. has applied the two-step proposal to a number of specific project categories, including power generation; industrial practices; methane capture; and land use, land use change and forestry (LULUCF). For each project category, a set of preliminary guidelines for determining additionality and quantifying emission credits is provided. These will be elaborated further as we go through the case studies in Appendix A.

> **Box 1. Sample Additionality Test.**
> Suppose we are developing a new coal plant. First, the reference case is developed by selecting a set of recent and comparable power plants. "Comparable" is interpreted as meaning same fuel; that is, other recent coal plants. Then the emission rates (CO_2/kWh) for each of the coal-fired plants in the reference case are quantified. Because no two plants are exactly the same, we can expect a range or distribution of emission rates. Then, a percentile test is applied to this distribution to quantify the eligibility threshold, referred to as X. For illustrative purposes, we will set X to 10 percent. At the 10th percentile the emission rate might, for example, correspond to 1.85 lbs CO_2/kWh. This value sets the performance threshold such that our coal project's emission rate would need to be less than 1.85 lbs CO_2/kWh to qualify as additional.

The EU's "Positive List" of Technologies

The EU has proposed that only projects based on a "positive list" of safe, environmentally sound, clean technology projects should be able to obtain credits.[4] The proposed positive list is presented in Table 1.

Table 1. The EU's Positive List of Technologies

Main Technology Categories	Individual Technologies
Renewables	Solar
	Wind
	Sustainable Biomass
	Geothermal heat and power
	Small-scale hydropower
	Wave and tidal power
	Ambient heat
	Biogas
Energy Efficiency	Advanced technologies for combined heat and power installations and gas fired power plants
	Significant improvements in existing energy production
	Advanced technologies for, and/or significant improvements in industrial processes, buildings, energy transmission, transportation and distribution
	More efficient and less polluting modes of mass and public transport (passenger and goods) and improvement or substitution of existing vehicles
Demand Side Management	Improvements in residential, commercial, transport and industrial energy consumption.

Few details are provided to clarify each category listed (renewables, energy efficiency, and demand side management) and the proposal provides no guidance on estimating baselines and calculating credits for projects that qualify under the positive list. However, the EU does specify that projects in the energy efficiency category, that involve fossil power plants, should be eligible for credit only if the following criteria are met:

- New Plants: If the plant has a minimum efficiency of 55 percent for plants larger than 300 MW and 52 percent for plants smaller than 300 MW

---

[4] European Commission, "Outcome of Climate Change Negotiations in Lyon, France, 4-15 September, 2000 (Press Release)," September 1, 2000, http://europa.eu.int/comm/environment/press/bio00172.htm

- Rehabilitation of Existing Plants: If the project introduces a technology change that leads to an increase in overall plant efficiency of at least five percent.

These criteria would, in effect, block most, if not all, current clean coal technologies from being used in new plant projects. It is, for example, conceivable that an IGCC system could be installed as a rehabilitative project that would lead to an increase in plant efficiency of more than five percent. However, the number of such projects may be very low. In addition to clean coal, nuclear power is absent from the specific list of qualifying technologies. The "positive list" supports the EU view that only low emissions sources should qualify for credits, reflecting a European trend of shifting towards alternative energies other than fossil fuels and nuclear energy.

The "positive list" has received significant support in Europe, particularly from the NGO community, which finds the "positive list" to be a very effective additionality test. Although the EU "positive list" was not specifically created for addressing additionality, in this report, we have taken the approach of evaluating the positive list according to its effectiveness as an additionality screen for projects.

The Full Technology Matrix

The technology matrix approach, modified and developed by the U.S. Department of Energy's National Energy Technology Laboratory (NETL) and illustrated in Table 2, consists of a selected list of greenhouse gas abating technologies that correspond with the sustainable development goals of a host country.[5] Like the official U.S. methodology, additionality and baseline determination under this approach takes place in two stages. First, a technology is subjected to an additionality test to determine whether it should be included in the matrix. This test would be based on factors such as the commercial viability and market penetration of the candidate technology. The test will be designed to ensure that only advanced, non-commercial technologies qualify for inclusion in the matrix.

Second, a stipulated benchmark will be developed for each approved technology based on the emissions performance of a selected group of counterfactual technologies. To qualify for credits, project developers would simply demonstrate that the proposed project technology is included in the matrix. The stipulated benchmark from the matrix would then be used to calculate the project's emission reductions. The emission baselines used for the individual technologies can be derived in several ways. In the energy sector for example, the baselines could be derived from the conventional technologies most likely to have been used, but for the project. Factors such as average heat rate, fuel mix, and best engineering practices would be included in the baseline. Moreover, technologies using a specific fuel type would be compared to a selection of projects and technologies using that same fuel. In this way, a baseline can be developed for each technology that

---

[5] As sustainable development criteria are likely to vary among countries most examples of the modified technology matrix are anticipated to be country-specific.

10

addresses the specific technology, policy, and economic issues impacting the reference scenario for that individual technology.

The modified technology matrix, however, does not cover all project activities that might satisfy a given set of additionality criteria. Because the technology matrix focuses on advanced, non-commercial technologies, additional projects utilizing conventional technologies will not qualify for credits under the technology matrix approach. To address this drawback the modified technology matrix approach should be used in conjunction with the project-specific approach. Under this combined methodology, the project-specific approach would be used to evaluate projects using conventional technologies while the modified technology matrix would work particularly well for projects using technologies that are not fully competitive or that have not yet been introduced in the individual host country.

Table 2. Example of a Portion of the Full Technology Matrix[6]

Countries Qualifying Technologies	India	China	Argentina	Philippines	Brazil
Coal-Fired IGCC	B_{IGCC-I}	B_{IGCC-C}	NA	B_{IGCC-P}	B_{IGCC-B}
Solid Oxide Fuel Cells (SOFC)	B_{SOFC-I}	B_{SOFC-C}	B_{SOFC-A}	B_{SOFC-P}	B_{SOFC-B}
Technology X	NQ	B_{X-C}	B_{X-A}	B_{X-P}	NA
Technology Y	B_{Y-I}	NA	B_{Y-A}	B_{Y-P}	B_{Y-B}

Notes: 1) B = Benchmark value for estimating project baseline emissions; 2) NQ = Technology choices that do not qualify as additional in a given country; and 3) NA = Not Applicable. Represents technology/country combinations that do not fit national sustainable development objectives.

The Hybrid Technology Matrix

The hybrid approach is based on a combination of the technology matrix's additionality test and the second step of the official U.S. proposal for baseline development. As we noted, the technology matrix additionality screen evaluates the commercial viability and market penetration of the candidate technology. If the project is found to utilize an advanced, non-commercial technology it will qualify for credits.

Once it has been established that the project qualifies, the credits will be determined. This will be accomplished by applying the U.S. methodology for benchmark development. This means that the reference scenario would be based on an average emission rate derived from recent and comparable activities in that same country or region, with fossil projects choosing from three categories of baselines 1) a fuel-specific average; 2) a fossil

---

[6] This table represents a hypothetical selection of host countries, technologies, and benchmarks that are included mainly for illustrative purposes.

average which would include coal, oil, and natural gas plants; and 3) a sector average, which would include fossil, hydropower, nuclear and renewable plants. The number of credits would then be determined by subtracting the project activity emissions from the average emissions of the reference scenario or, in the case of fossil projects, one of the three baseline categories.

The Case Studies

Forty case studies have been developed, covering the electricity, industrial, transportation, land use, and residential and commercial sectors. In developing the case studies, the objective has been to cover a variety of plausible projects. Only by applying the four approaches to as wide of a variety of projects as possible can these approaches be tested for the full spectrum of situations likely to arise under a future carbon mitigation regime. Our guide, in this attempt to cover as many project types as possible, has been the database developed under the Energy Information Administration's Form EIA-1605 survey. As discussed in Chapter 1, The EIA-1605 is a voluntary program ("Voluntary Reporting of Greenhouse Gases"), in which corporations, organizations, and other entities report information on their greenhouse gas emission reduction efforts. No credits or other incentives are awarded for the emission reductions reported on Form EIA-1605. Nonetheless, the program has attracted significant participation. In 1998 (the last year for which the data have been made public), 187 U.S. companies and other organizations reported on a total of 1507 emission reduction projects. The 187 participants included electric utilities and non-utility generators accounting for the majority of total U.S. generating capacity. In addition, many other major (as well as smaller) U.S. companies report on Form EIA-1605, including, e.g., General Motors, IBM, Dow, and Johnson and Johnson.

Voluntary participants in the program are driven to report by a variety of objectives. For example, other government programs, including the U.S. Department of Energy's Climate Challenge and Climate Wise, have adopted the Form EIA-1605 as their reporting instrument. Participants in Climate Challenge and Climate Wise voluntarily pledge to reduce their emissions by specified amounts. These participants report progress towards achieving their emission reduction goals on Form EIA-1605. Other organizations, not involved in Climate Challenge or Climate Wise, are driven to report on Form EIA-1605 by their desire to publicize their environmental efforts.

But certainly one of the primary objectives of many of the participants in the EIA-1605 program is to lay claim to any credits that might, at some point in the future, be awarded to voluntary emission reduction efforts undertaken in the past and present. In effect the Form EIA-1605 serves as an official registry of claims to future credits for past and current emission reductions. There is no guarantee that future legislation will ever be passed to award emission reduction efforts reported on Form EIA-1605. Nor is there any guarantee that, even if such legislation is passed, any of the claims registered under the Form EIA-1605 program will meet the qualification criteria for receiving credits. Nonetheless, the possibility that such credits might be awarded, at some unknown future date, has induced many corporations to participate in the program.

Thus, the Form EIA-1605 database provides a reasonably good indication of the types of emission reduction projects that are being claimed as credit worthy within the United States. Whether credits for these same types of projects might also be claimed under some future carbon mitigation program is, of course, unknown. However, the Form EIA-1605 would appear to provide at least a reasonable guide in the development of *plausible* projects that *might* be undertaken under a future program. In addition, in assessing project assessment methodologies, it seems prudent to test these methodologies against as wide a spectrum of *plausible* projects as possible. If such a full test is *not* performed, then there exists the very real possibility that the chosen methodology will prove unworkable when applied to certain classes of projects. The applicability of the potential candidate methodologies to the various possible project types is best determined *prior* to program implementation using hypothetical case studies, rather than during program implementation using actual projects as the test cases.

In reviewing the Form EIA-1605 database, one is struck by the sheer variety of projects claiming emission reductions. Heat rate improvements, fuel switching, cogeneration, high-efficiency transformers, reconductoring, landfill gas recovery, coal mine methane recovery, renewables, afforestation, reforestation, urban forestry, lighting and lighting controls, HVAC, building shell improvement, motor and motor drive, transportation, halogenated substance reduction projects, recycling, and methane recovery from wastewater treatment are just a few examples of the types of projects reported in 1998.

Table 3 provides a more complete (though not comprehensive) list of the projects reported on Form EIA-1605 in 1998, by sector. The table shows the main project type categories by various sectors, along with the number of projects of each type reported in 1998. The table also shows the case studies designed to represent important project type categories that are included in Appendix A. Case studies that have already been developed are identified in Table 3 by their Project Number ID; for example, ES1 identifies project number 1 in the electricity supply sector. As Table 3 shows, case studies have been developed for many, but by no means all, of the project type categories currently being reported on Form EIA-1605. Although this report falls short of a comprehensive coverage of all potential project types, it is believed that many of the most important categories are covered. Further case study development and analysis is no doubt warranted.

Table 4 provides a more detailed summary of the case studies included in Appendix A. In this table, and in the appendix, the case studies are organized by sector; however, we have used a different sectoral organization than that utilized by the EIA-1605 program and shown in Table 3. As Table 4 indicates, 11 case studies have been developed for the electricity sector, 13 for the industrial sector, 9 for the transportation sector, 2 for the forestry sector, 3 for the residential sector, 2 for the commercial sector, and 2 for the forestry sector. Individual case studies run the gamut from IGCC, through wind power,

Table 3. Types of Projects Reported on Form EIA-1605 in 1998

Sector	Type of Project	Number of Projects*	Corresponding Case Study(ies)
Electric Power (Including Energy End-Use)	Heat Rate Improvement	166	ES2
	Availability Improvement	29	
	Fuel Switching	42	ES3
	Increase in Low-Emitting Capacity	78	ES1, ES4-ES9, ES11
	Decrease in High-Emitting Capacity	9	
	Cogeneration	17	IS2
	Dispatching Changes	7	
	Zero/Low Emission Power Purchase	6	
	Other Generation	13	
	High-Efficiency Transformers	42	
	Reconductoring	28	ES10
	Distribution Voltage Upgrade	28	
	Other Transmission and Distribution	18	
	Equipment and Appliances	89	
	Lighting and Lighting Control	131	RS2
	Load Control	45	
	Heating, Ventilation and Air Conditioning	80	CS1, CS2
	Building Shell Improvement	54	IS8, RS1, RS3, CS1
	Motor and Motor Drive	48	IS3, IS4, CS2
	End-Use Fuel Switching	16	
	Industrial Power Systems	1	
	Urban Forestry	7	
	Other End Use	33	
Alternative Energy Providers	Landfill Gas Recovery for Flaring and Energy	5	
	Landfill Gas Recovery for Energy	54	
	Landfill Gas Recovery for Flaring	20	IS12
	Source Reduction at Landfills	1	
	Coal Mine Methane	12	IS11
	Biomass	5	
	Other Renewables	6	ES6
Agriculture and Forestry	Afforestation	8	LU2
	Reforestation	97	
	Sequestration from Urban Forestry	1	
	Forest Preservation	4	LU1
	Woody Biomass Production and Other Agroforestry	1	
Industrial	Oil, and Natural Gas Systems and Coal Mining: Methane	4	IS11, IS13
	Cogeneration	1	IS2
	Energy End Use	79	IS3-IS5, IS8-IS10
	Transportation	6	TS1-TS9
	Carbon Sequestration	1	ES11, LU1, LU2
	Halogenated Substances	11	IS6
	Other	10	

*Some projects are counted in more than one project type category.

hydropower, district heating, cogeneration, PFC reductions, and CNG vehicles. Although electricity supply projects tend to dominate the reporting under the Form EIA-1605 program (see Table 3), we have chosen to develop approximately the same number of industrial and transportation sector projects as electricity supply projects.

Thus far, in the development of both the technology matrix and official U.S. methodologies, electric power projects have received significantly more attention than projects in other sectors. This is reflected in the official U.S. approach, which has been developed in much greater detail for electricity supply projects than for other project types. As a result of this bias towards electricity supply, it is likely that both the official U.S. approach and the technology matrix approach are more well adapted to electricity supply projects than to other project types. Therefore, in this case study analysis, we have shifted the emphasis somewhat towards other sectors, in order to test the methodologies in applications which, although receiving less attention thus far, are nonetheless plausible under future carbon mitigation regimes.

Case Study Development and Analysis Methodology

All of the case studies identified in Table 4 are fictitious. In addition, most (although not all) of the "data" utilized in the case studies are fictitious. The use of hypothetical projects, with fictitious data, significantly reduced the amount of time required to develop each case study. This in turn enabled the development of a large number and variety of case studies—an important objective of the analysis, given the desire to test the methodologies under the full spectrum of plausible scenarios. Had an attempt been made to obtain actual data for the case studies, the data collection effort would have drastically

Table 4. The Case Studies

Sector	Project ID Number	Country	Project Title
Electricity	ES1	India	IGCC Power Plant
	ES2	India	Heat Rate Improvement
	ES3	India	Fuel Switching
	ES4	India	Natural Gas Combine Cycle
	ES5	India	Gas Turbine Plant
	ES6	India	Wind Power
	ES7	Kazakhstan	IGCC in Kazakhstan
	ES8	Tajikistan	Hydropower
	ES9	India	Distributed Generation: Fuel Cells
	ES10	China	Transmission Capacity Expansion
	ES11	India	Carbon Sequestration for IGCC Plant
Industrial	IS1	Azerbaijan	Installation of District Heating System
	IS2	Kazakhstan	Cogeneration at Food Processing Plant
	IS3	Argentina	Variable Frequency Drives
	IS4	Brazil	Retrofit of Energy Efficient Motors
	IS5	China	Coke Oven Underfiring Rate Improvement
	IS6	Tajikistan	PFC Reductions at Aluminum Plant
	IS7	China	Coal Ash Utilization
	IS8	Chile	Building Insulation Improvement
	IS9	Jordan	Highly Efficient Fertilizer Complex
	IS10	China	Industrial Boiler Shutdown
	IS11	South Africa	Coal Mine Methane Recovery
	IS12	Argentina	Landfill Gas Flaring
	IS13	Kazakhstan	Recovery of Associated Natural Gas
Transportation	TS1	India	Dedicated CNG Taxis

Sector	Project ID Number	Country	Project Title
	TS2	India	New Gasoline-Fueled Taxis
	TS3	China	Aluminum Rail Cars for Efficient Coal Transport
	TS4	S. Africa	Clean Diesel in Transit Buses
	TS5	Mexico	Electric Vehicles in Mexico City
	TS6	Thailand	Smart Toll System
	TS7	Ukraine	46 New Conventional Diesel Buses
	TS8	India	New Two-Wheelers
	TS9	Brazil	Improving Road Infrastructure
Land Use	LU1	Mexico	Forest Protection and Management
	LU2	Russian Federation	Afforestation of Marginal Agricultural Land in Russia
Residential	RS1	South Africa	Construction of Energy-Efficient Homes in South Africa
	RS2	Mexico	Sale of High-Efficiency Light Bulbs for Homes
	RS3	Russian Federation	Energy Efficiency of Seven Apartment Buildings
Commercial	CS1	Philippines	Energy Efficiency and Conservation Measures in Commercial Buildings
	CS2	Indonesia	Motor Replacement Project in Commercial Office Buildings in Jakarta

reduced the amount of time available for case study development and analysis. Furthermore, in many cases it would likely have proved impossible to obtain the required data. In fact, as shall be discussed in the next chapter, data availability is a major concern for all four methodologies.

Further, had an attempt been made to use actual rather than hypothetical projects, key issues might have been missed. The case study development process was designed, in part, to illustrate certain issues and concerns already recognized as potential problem areas. Some, though not all, of the case studies were specifically developed to highlight these problem areas. Other case studies were designed simply to represent project types that are likely to be undertaken in significant numbers. The case study analyses by no means served merely to illustrate pre-recognized issues; many new issues were identified during the case study development and analysis effort.

Each case study follows a standard format or template. We will use Project Number ES1, in the appendix, to illustrate this template. The first page of the case study for ES1 provides a basic description of the project, along with the information and data needed to analyze the project. As indicated in the project description, ES1 involves the construction of a new integrated gasification combined cycle (IGCC) power plant in India. The project description for ES1, and for all the other case studies, provides some basic technical information on each project, followed by a description of the financing arrangements for the project. The project description is followed by a discussion of the project's additionality status. In this discussion, we assume that we possess an omniscient knowledge of the motives of the project developers. Thus, in the case of ES1, we can discern that the developers chose to build the IGCC plant in order to obtain emission reduction credits. This omniscient knowledge of the motives of the project developers in turn allows us to determine, definitively, whether or not the project is in fact a free rider. Such a definitive determination of motives and hence additionality is never possible in the case of real projects. This is another important advantage of the hypothetical case study approach. Our use of fictional rather than actual projects enables us to adopt a position akin to that of an omniscient narrator of a novel. Just as such a narrator can describe the inner thoughts of the novel's characters, we can describe the motives of our fictional project developers.

The section on project additionality is followed by two sections providing the quantitative data needed to apply the four methods to the project. First, the "Project Emissions" section presents the emissions rate for the project, along with the backup data used to compute the emissions rate. In the case of project ES1, the emissions rate is specified in units of pounds gas per kilowatt-hour, but the units vary across case studies depending on the specifics of each project. For projects involving modifications to existing facilities, the "Project Emissions" section also presents the data needed to compute the facilities' emissions rates prior to the projects. The "Project Benchmark Data" section presents the data needed to compute the "threshold" emissions rate under the official U.S. approach, as well as the project baselines under all approaches (with the exception of the EU approach, which does not address baseline development). In the case of ES1, actual, relevant data on five recently built coal-fired power plants were available and hence were used in the "Project Benchmark Data" section. However, in most cases the benchmark data presented in the case studies is fictional, although an effort has been made to ensure that these data, as well as the project emissions rate data, are plausible. The project benchmark data for ES1 includes plant capacities, heat rates, and emissions rates; however, the specific data provided varies depending on the analysis requirements for each case study.

The first page of case study ES1, presenting the basic project description along with the information and data required for the project analysis, is followed on the second page by the "Project Analysis Table." The Project Analysis Table presents the analysis of the project for each of the four methodologies. The table follows the same format for all case studies. Each of the four columns corresponds to one of the four methods. In the first row of the table, the project is tested, according to the rules established under each of the four methods, to determine whether or not it qualifies for credits. In the case of ES1, for example, we find that the project would qualify for credits under the official U.S. approach, as long as the value of "X" in the percentile threshold test is set above 20. Below the 20th percentile, the number of "recent, comparable" power plants in India is insufficient to support the threshold test. However, it is relatively safe to assume that, if more power plants could be included in the benchmark group, e.g., by expanding the geographic area covered beyond the borders of India, the project's emission rate would still fall below the threshold. The project thus appears to qualify for credits regardless of the value of X. Similarly, the project qualifies for credits under both the full and hybrid technology matrix approaches, because IGCC is an advanced, non-commercial technology. However, the project fails to qualify for credits under the EU's positive list, because it is a fossil fuel project with an efficiency below 55 percent.

In the second row of the table, the question is asked whether the project has been correctly identified as a free rider or as additional under each of the four approaches. Since it has already been established (under the "Project Additionality" section) that ES1 is truly additional, the second row indicates that all but one of the approaches have correctly qualified the project as additional. The sole exception is the EU approach.

In the third row of the table, the number of credits to be awarded to the project under each approach is computed. The benchmark, under the U.S. approach, would be set equal to the weighted average emissions rate of the five recently opened coal-fired power plants (as provided in the "Project Benchmark Data" section). These five plants constitute the "reference scenario" under the U.S. approach. The same benchmark would be used for the hybrid technology matrix, because this approach uses the same methodology to compute the quantity of credits to be awarded as the official U.S. approach. In this particular example, the benchmark under the full technology matrix is the arithmetic average emissions rate for the same five coal-fired power plants. The same plants are used to define both the "reference scenario" for the U.S. approach and the technology matrix's "model counterfactual," because in this particular case the "model counterfactual" for coal-fired IGCC is represented by recently-opened, conventional coal-fired power plants.[7] However, in other cases, the "model counterfactual" under the full technology matrix differs from the "reference scenario" under the U.S. approach. One of the important features of the full technology matrix is that it requires the specification of a separate model counterfactual for each technology. In contrast, the U.S. approach utilizes only five reference scenarios for the electricity sector: a sector-wide reference scenario, a fossil reference scenario, and three reference scenarios for the three main fuel types (coal, oil, and gas).

In the case of ES1, as in all of the case studies, the number of credits awarded cannot be determined under the EU proposal. This proposal does not specify a procedure for computing the quantity of credits. Therefore, for all of the case studies, the number of credits awarded under the EU proposal is specified as "not applicable" in the Project Analysis Table.

The fourth row of the Project Analysis Table provides an estimate of the error in the credits awarded, when possible. An error estimate is not possible based on the information provided in the case of ES1. However, for some of the other case studies error estimates can be made.

Finally, the third page of ES1 presents the "Methodology Assessment." This section of the case studies discusses the key "lessons learned" from the project analyses. In developing the Project Analysis Tables, the four methods are simply applied to the projects without comment or critique. However, the purpose of the "Methodology Assessment" is to provide a critical analysis of the applicability of the four methods to

---

[7] See the SAIC report "Developing the Technology Matrix for India and Ukraine" (August 2000). On pages 48 through 61, this report defines the model counterfactual for IGCC technology in India.

the case study. For example, in the case of ES1, the EU proposal's qualification test is found wanting, because it rejects this truly additional project. In addition, the relative difficulty of meeting the data requirements under the U.S. proposal is discussed. In general, the "Methodology Assessment" focuses on two or three important problem areas or issues illustrated by the case study.

4. SUMMARY AND LESSONS LEARNED

This chapter presents a summary of the key issues and "lessons learned" from the individual case studies. Included in this chapter are recommendations for improving the different proposals. The chapter will conclude with a discussion of NETL's technology matrix approach as the technical solution to many of the problems posed by the requirements of project evaluation.

Summary of Case Study Findings

The case studies include 11 projects in the electricity sector, 13 projects in the industrial sector, 9 transportation projects, 2 forestry projects, 3 residential projects, and 2 commercial projects. Table 5 summarizes the findings of the case study analyses in terms of whether each specific baseline development approach correctly identified the projects as additional or free riders.

Table 5. Case Study Results

	Project Information				Case Study Result: Is the Project Correctly Identified as Additional or Free Rider?			
ID	Country	Title	Addi-tional	Free Rider	US Approach	EU Positive List	Full Technology Matrix	Hybrid Technology Matrix
Electricity Generation								
ES1	India	IGCC Power Plant	✔		Yes	No	Yes	Yes
ES2	India	Heat Rate Improvement		✔	Depends on X	Yes	Yes	Yes
ES3	India	Fuel Switching		✔	Depends on X	Yes	Yes	Yes
ES4	India	Natural Gas Combined Cycle		✔	Yes	No	No	No
ES5	India	Gas Turbine Plant	✔		No	No	No	No
ES6	India	Wind Power	✔		Yes	Yes	Yes	Yes
ES7	Kazakhstan	IGCC Power Plant	✔		Yes	No	Yes	Yes
ES8	Tajikistan	Hydropower		✔	No	Yes	Yes	Yes
ES9	India	Distributed Generation: Fuel Cells	✔		Yes	No	Yes	Yes
ES10	China	Transmission Capacity Expansion		✔	No	Indeter-minate	Yes	Yes
ES11	India	Carbon Sequestration Technology for an IGCC Power Plant	✔		Yes	No	Yes	Yes
Industrial Sector								
IS1	Azerbaijan	Installation of District Heating System		✔	Yes	Indeter-minate	Yes	Yes
IS2	Kazakhstan	Cogeneration at Food Processing Plant		✔	Yes	Indeter-minate	Yes	Yes
IS3	Argentina	Variable Frequency Drives	✔		Inde-terminate	Yes	No	No
IS4	Brazil	Retrofit of Energy Efficient Motors		✔	No	Indeter-minate	Yes	Yes
IS5	China	Coke Oven Underfiring Rate Improvement	✔		Yes	Yes	Yes	Yes
IS6	Tajikistan	PFC Reductions at Aluminum Plant	✔		Yes	No	Yes	Yes
IS7	China	Coal Ash Utilization	✔		Yes	Indeter-minate	No	No
IS8	Chile	Building Insulation Improvement	✔		No	Indeter-minate	Inde-terminate	Inde-terminate
IS9	Jordan	Highly Efficient Fertilizer Complex		✔	Yes	Indeter-minate	Inde-terminate	Inde-terminate

Project Information					Case Study Result: Is the Project Correctly Identified as Additional or Free Rider?			
ID	Country	Title	Additional	Free Rider	US Approach	EU Positive List	Full Technology Matrix	Hybrid Technology Matrix
IS10	China	Industrial Boiler Shutdown		✔	Indeterminate	Yes	Yes	Yes
IS11	South Africa	Coal Mine Methane Recovery	✔		Indeterminate	Yes	No	No
IS12	Argentina	Landfill Gas Flaring		✔	Indeterminate	Indeterminate	No	No
IS13	Kazakhstan	Recovery of Associated Natural Gas		✔	Indeterminate	Indeterminate	Yes	Yes
Transportation								
TS1	India	Dedicated CNG Taxis	✔		Yes	Yes	Yes	Yes
TS2	India	New Gasoline-Fueled Taxis		✔	No	Indeterminate	Yes	Yes
TS3	China	Aluminum Rail Cars for Efficient Coal Transport		✔	Indeterminate	No	No	No
TS4	South Africa	Clean Diesel in Transit Buses		✔	No	No	No	No
TS5	Mexico	Electric Vehicles in Mexico City	✔		Yes	Yes	Yes	Yes
TS6	Thailand	Smart Toll System	✔		Indeterminate	Yes	Yes	Yes
TS7	Ukraine	46 New Conventional Diesel Buses	✔		Yes	Yes	No	No
TS8	India	New Two-Wheelers	✔		Depends on X	Yes	Yes	Yes
TS9	Brazil	Improving Road Infrastructure	✔		Indeterminate	Yes	Indeterminate	Indeterminate
Land Use/Forestry								
LU1	Mexico	Forest Protection and Management	✔		Yes	No	No	No
LU2	Russian Federation	Afforestation of Marginal Agricultural Land	✔		No	No	Indeterminate	Indeterminate
Residential								
RS1	South Africa	Construction of Energy-Efficient Homes in South Africa	✔		Yes	Yes	No	No
RS2	Mexico	Sale of High-Efficiency Light Bulbs for Homes		✔	Indeterminate	No	No	No
RS3	Russian Federation	Energy Efficiency of Seven Apartment Buildings	✔		Yes	Yes	No	No
Commercial								
CS1	Philippines	Energy Efficiency and Conservation Measures in Commercial Buildings	✔		Yes	Yes	Indeterminate	Indeterminate
CS2	Indonesia	Motor Replacement Project in Commercial Office Buildings in Jakarta	✔		Yes	Yes	Yes	Yes

Lessons Learned

The case study analysis reveals several interesting and sometimes surprising results. To varying degrees, all of the project evaluation approaches demonstrate the capacity to allow non-additional, free rider projects to obtain credits, and to disqualify truly additional projects. Often, these qualification errors differ among the approaches, making generalizations regarding project types difficult. Moreover, some project types simply fail to be suitable for evaluation under standardized approaches. In the following discussion, we provide a summary of the major findings and recommendations derived from our analysis. We begin with a summary of issues that are relevant for all of the four project development approaches. Then we examine issues that are more specific to each baseline methodology.

Classification and Misclassification of Projects

The case studies reveal that, at various points, all four standardized project development approaches fail to correctly identify some projects as additional or free riders. Table 6 summarizes the classification trends of each baseline proposal. Application of the U.S. approach resulted in a higher number of indeterminate responses for the additionality test compared to the other approaches. The EU and the technology matrix approaches tend to fail by mis-classifying truly additional projects as free riders. Moreover, the EU positive list only correctly identified 18 out of 40 projects examined, highlighting the difficulties faced by the EU positive list as an applicable project evaluation screen for many of the different technology and project types presented. The additionality classification results of the full technology matrix and the hybrid technology matrix are similar. This is because both of these approaches apply the same additionality test. Differences in these two approaches appear during the estimation of credits.

Only projects that reduce greenhouse gas emissions beyond what is expected as part of the projected business-as-usual scenario should be viewed as credit-worthy. Project development approaches must therefore be capable of screening out those free rider projects that have already been accounted for as business-as-usual. The project evaluation approaches used should focus on conservative and stringent baseline estimation procedures that minimize the level of error of the credits awarded. With this goal in mind, it appears that the technology matrix approaches are slightly better qualified for evaluating projects for a greenhouse gas offset programs. However, when making this conclusion, it should be noted that our analysis is based on an examination of hypothetical case studies.

Table 6. Summary of Classification Trends of Baseline Approaches

Case Study Result	US Approach	EU Positive List	Full Technology Matrix	Hybrid Technology Matrix
Projects correctly identified as additional or free riders	20	18	22	22
Free rider projects awarded credit	5	4	5	5
Additional projects not awarded credit	3	8	8	8
Indeterminate Projects	9	10	5	5
Unclear. Classification depends on percentile threshold (X)	3	N/A	N/A	N/A

The Necessity of a Backup Methodology

Each of the four baseline approaches encountered situations where the methodologies could not be applied and the free rider/additionality status of the project could not be determined. The EU positive list proved to encounter the most difficulties. All together ten projects were indeterminate under the EU approach. This failure stems primarily from the imprecise and vague language used in the EU proposal. For example, the positive list states that energy efficiency projects that rely on "significant improvements in existing energy production" will receive credits. It is impossible to decide, in an objective manner, what constitutes "significant" improvements.

The U.S. approach encountered nine situations where the methodology could not classify a project, and three situations in which the classification would depend on the value chosen for X. For example, project IS3 (Variable Frequency Drives in Argentina) could not be evaluated effectively because it involves energy conservation rather than an improvement in efficiency. Due to a lack of guidance for the treatment of energy conservation projects within the U.S. approach, an appropriate reference scenario could not be identified for this project. IS3 illustrates that until now, the developers of the U.S. approach have focused mainly on the procedures for evaluating fuel switching and energy efficiency projects. Default or back-up guidelines still need to be established for other types of emission reduction activities. In the case of project IS10, involving an industrial boiler shutdown in China, the additionality of the project could not be determined because the U.S. does not specify whether this type of project should be evaluated based on a comparison of energy consumption data or emission rates. In this particular case, each comparison would provide a different conclusion regarding the additionality of the project.

The technology matrix approaches also encountered difficulties when applied to five of the case studies. For example, project IS9 (Highly Efficient Fertilizer Complex in Jordan) could not be tested for additionality because it involves an advanced technology application for only one component of the project. In addition to two indeterminate classifications of projects in the industrial sector, the technology matrix also resulted in indeterminate classifications of a land use project, a commercial, and a transportation project.

Given that each methodology encountered projects that it could not handle, the need for a back up project evaluation approach is apparent. The technology matrix approaches specifically recommend using the project-specific approach in those instances where the technology matrix alone would either disqualify the project or lead to an indeterminate additionality classification. However, neither the U.S. proposal nor the EU positive list incorporate such back up procedures. Unofficially, the promoters of these two baseline methodologies may understand the need for a back up. However, such an escape clause should be clarified up-front and procedures for applying the default methodology should be specified.

As noted earlier, the technology matrix approaches rely on the project-specific approach as a fallback procedure. The project-specific approach could also serve as a suitable fallback procedure for both the U.S. approach and the EU positive list. However, it should be mentioned that although the project-specific approach is useful as a fallback methodology, as it can be effectively applied to most types of projects, it also has a major drawback. The project-specific approach tends to result in higher transaction costs that discourage smaller-sized projects. This may discourage small-scale projects that are truly additional, such as the gas turbine project in India (ES5) (which failed to qualify for credits under all four of the baseline methodologies).

Deficiencies in Data Availability

The use of standardized project evaluation procedures increases the need for sufficient and reliable data, both for determining the benchmarks and, in some instances, for undertaking the additionality test. However, as illustrated in the case studies, the required data is unavailable in many developing countries and in some cases, the necessary data simply cannot be derived because there are no comparable activities upon which a benchmark can be based. These data issues cut across all methodologies, but are of greater significance for the U.S. methodology because of its relatively extensive data requirements. The following is a list of major data issues raised by the case studies.

Lack of Data: Many developing countries currently lack the necessary data to support emission reduction accounting and do not have the funding to support data collection efforts. These data deficiencies impede baseline development under both the U.S. and technology matrix approaches, because each of these approaches requires sector-wide data for benchmark development. The issue of limited data availability is particularly apparent for projects in the industrial and transportation sectors, where the large variation in activity type, size, and project participants complicates uniform data collection. For the industrial sector, examples of data problems include locating matching plants, lack of private sector reporting procedures, and restricted access to data due to private sector confidentiality. In the transportation sector, it is difficult to obtain data describing the emission performance of particular vehicle types and models as they age. For the power sector, a major problem has been the lack of information on fuel consumption. In most instances, we have used fictitious data to evaluate the case studies. Previous efforts at collecting baseline data for the power and transportation sectors in India and Ukraine have proved to be very costly and time consuming. Local consultants have been working with NETL over the past year to collect and develop data. In spite of these concentrated activities, some of the required data is still unavailable – mostly because it simply does not exist in the respective countries. Thus, to reduce cost and increase the number and variety of case studies fictitious data were used in those instances where real-life data was not readily available.

Because of the deficiencies in data availability, it will likely be necessary to develop regional or global default benchmarks. In many cases, such global benchmarks may suffice. However, the situation posed in project ES8, involving construction of a hydropower plant in Tajikistan, indicates the potential problem of relying on regional data. In this particular case study, data is unavailable on either a national or a regional basis to support the development of benchmarks. Thus, it is impossible to develop the sector average that is needed to calculate credits. This raises two problems. First, the U.S. proposal requires a comparison to "recent" facilities. However, there may not be any recently constructed plants in any former Soviet Union (FSU) countries from which to construct the data. Secondly, if recent plant data from other FSU countries were available, it would be difficult to utilize that fossil fuel dominated data to accurately reflect the sector average in hydro-dominated Tajikistan.

Although the lack of adequate data affects all methodologies, it has the largest impact on the U.S. approach. Under the U.S. methodology, sector-wide data is required both to develop the percentile threshold for evaluating additionality and to derive the benchmark. Moreover, the threshold test requires data, not only for deriving a weighted average, but also for developing a distribution that is sufficiently defined to establish "X." For the technology matrix approaches, on the other hand, sector-wide data is used only for estimating the benchmark.

Unless these data problems are addressed prior to the implementation of an emission reduction program, project developers will be faced with the expense of collecting the data themselves. This would significantly add to project costs. Moreover, if project developers are left to collect the data, the objectivity and accuracy of the data could be called into question. Alleviating some of the difficulties in obtaining adequate data will require institutional capacity building in potential host countries to build the institutions and resources necessary to collect and process required information.

Data Requirements for Representing "Recent and Comparable" Activities: Another issue adds to the data problems associated with the U.S. approach; that is, the criterion that projects must be compared to "recent and comparable" activities. As illustrated in the IGCC power plant case study in Kazakhstan (ES7), a problem arises when the host country, and neighboring countries, lack the recent, comparable facilities needed to support the development of country-specific or regional percentile tests and benchmarks. In this particular case, no new power plants have been built in Kazakhstan or surrounding FSU countries in recent years. This problem of non-existent recent facilities is likely to be most pronounced in the FSU. Within the FSU new capacity projects have been rare in recent years due to the economic collapse resulting from the breakup of the Soviet Union. Consequently, there is a general lack of recently built facilities, which can serve as a basis of comparison for new emission reduction projects. Presumably, this problem will have to be addressed by establishing global or regional default data. However, the procedures for establishing default data should be clarified under the U.S. approach to encourage transparency.

In many other cases, some data is available; however, the amount of data provided is not substantial enough to result in an effective additionality test. For example, under the U.S. approach, data sufficient to define a percentile distribution (such as more than five data points if a threshold of $X<20^{th}$ percentile is to be used) are required. Yet in India, for example, there are only five coal-fired power plants built within the last five years which possess credible data that can be used for benchmark development. In order to establish an adequate distribution for the percentile test, it would thus be necessary to include regional data. Again, it becomes apparent that the U.S. methodology requires substantially more data than the technology matrix approaches to serve as an effective project evaluation tool.

Procedures for Data Treatment: Often the data obtained for benchmark and percentile threshold development includes data that is suspect or flawed. For example, in the first two electricity projects (ES1 and ES2) in India, data from two power plants were removed because the credibility of the data provided was suspicious. In other cases, the unique characteristics of each individual facility may make it impossible to identify *any* comparable facilities. This issue was raised in the analysis of case study IS8 (Building Insulation Improvement in Chile). In this instance, two of the potential comparable facilities examined are located in two different parts of the host country (latitude 37° South versus latitude 18° South) and possibly face very different climate conditions. In addition, there is a large difference in the energy used for heating by each facility (50,000 $Btus/ft^2$-yr versus 5,000 $Btus/ft^2$-yr). An even larger problem arose in the case of project IS9 (Highly Efficient Fertilizer Complex in Jordan). In this particular case, no similar industrial complexes had been built in the host country or region, prompting the use of global data. This scenario is likely to be encountered in many host countries that have less developed industrial sectors. Moreover, for projects (such as IS9) involving large, complex industrial facilities, that produce numerous outputs in variable proportions, it may be impossible to find comparable data (even at the global level).

The fact that comparable data may be completely non-existent, or that a project may qualify under one set of data points but not under another, is a major concern. To alleviate this problem, sample protocols for collecting, processing and presenting data should be developed to facilitate standardization, verification of credits, and replication of baselines.

The EU Positive List

The EU positive list encountered a host of problems as it was applied to the different case studies. These problems make manifest the fact that this approach is relatively undeveloped compared to the other approaches. In its current form, it does not represent an operational and effective baseline test. However, because the EU positive list appears very simple to apply, it is attractive to policy makers interested in promoting a transparent and environmentally stringent solution. As it turns out, the lack of specificity and clarity would probably prolong the evaluation process and increase the transaction costs associated with project development.

In the following discussion, we have summarized the major problems that were raised by applying the case studies to the EU positive list.

The first and most substantial problem associated with the EU positive list is the lack of clarity in the definition of qualifying technologies and processes. Several times, classification of projects was hampered due to unclear criteria. For example, the EU positive list allows for energy efficiency projects that significantly improve energy transmission. However, in the absence of a definition of the term "significantly," it cannot be determined whether or not projects, such as ES10 (Electricity Transmission Capacity Expansion in China), would qualify as additional. Similarly, the criteria fail when examining co-generation projects (see project IS2, Co-generation at Food Processing Plant in Kazakhstan). The EU positive list allows "advanced technologies" to qualify. However, it is unclear whether this means that all co-generators qualify because co-generation is itself considered an advanced technology, or whether only co-generators using "advanced technologies" will qualify. Similarly, the wording of the criteria for transportation projects is vague. The EU proposal states that projects deploying "more efficient and less polluting modes" of transportation qualify for credits. Thus far, the positive list offers no objective, quantifiable definition of terms such as "significantly," "more efficient," and less polluting," making it impossible to determine whether or not projects would qualify as additional.

Secondly, the categories in the positive list are developed in a fashion that allows some projects to qualify under more than one main category. Project IS4 (Retrofit of Energy Efficient Motors in Brazil) and project IS7 (Coal Ash Utilization in China) both fit under the categories of energy efficiency and demand side management. The fact that a single project could potentially fall into two separate categories under the EU positive list, resulting in different determinations of additionality, is clearly problematic.

A third drawback of the positive list is that it addresses only energy-related projects, thus in effect automatically disqualifying many classes of projects that reduce greenhouse gases other than carbon dioxide. For example, in the case of project IS6 (PFC Reductions at an Aluminum Plant in Tajikistan), emissions of perfluorocarbons (PFCs) are reduced by implementing a computerized process control system. Because project IS6 improves process efficiency, not energy efficiency, it does not fit in any of the qualifying technology categories included in the positive list. Yet non-energy related projects such as IS6 often provide a highly cost effective method for reaching global emission reduction goals.

Fourth, because of the failure to include clean coal activities on the positive list, the EU approach excludes certain types of large-scale projects, such as IGCC. The goal under the EU approach is to push developing countries away from coal and other fossil fuels, and towards renewables. However, it is unlikely that, for any specific project, the potential options would come down to a choice between IGCC or other clean coal technologies, and renewables. These two classes of technologies are designed to serve very different applications. Clean coal projects, such as IGCC, typically fill large capacity, baseload generation needs. Renewable technologies such as solar and wind cannot be used for such large-capacity, baseload applications. Thus clean coal projects such as IGCC, if disqualified under the EU positive list test, are not likely to be replaced by renewable energy. Instead, they will be replaced by conventional fossil fuel sources.

In short, the goal of promoting renewables by excluding clean coal is probably unattainable, and will likely lead only to an increased reliance on conventional coal.

Finally, the EU positive list appears to inadvertently exclude certain technologies. For example, project ES9 in India involving distributed electricity generation based on fuel cells fails to qualify because fuel cell technology is not on the positive list. Given the EU's emphasis on clean and renewable energy development, fuel cells would be an obvious candidate for inclusion in the positive list. Conversely, the EU approach will always qualify other types of technologies and projects once they have been added to the list. For instance, all the residential and commercial projects examined automatically qualified because the EU list allows for any improvements in residential, commercial, transportation, and industrial energy consumption, without adding any additional criteria to screen out business-as-usual projects.

This raises another important point. The positive list approach does not define procedures for adding new technologies or removing old technologies from the list. Any standardized approach must include a methodology that accounts for technological change over time.

Altogether, a number of improvements can be undertaken to develop the EU positive list into a more comprehensive baseline approach. These improvements include:

1) *Clarifying definitions of qualifying technologies and processes and specifying vague language by assigning quantifiable values to each criterion.* In particular, these specifications should emphasize language that is exclusive and prevents projects from fitting into more than one category. Out of 22 case studies, seven projects were categorized as indeterminate under the positive list, mainly due to vague language. Improved procedures for qualifying projects will significantly reduce the classification failures of the positive list. Among other steps, improvements should include an explicit list of specific technologies and processes that qualify for credits. Examples of entries on this list might include, e.g., "CNG vehicles," "hybrid vehicles," etc., in place of vague entries such as "more efficient and less polluting modes of mass and public transport."

2) *Developing a methodology for quantifying emission reduction credits.* Currently the positive list does not include a methodology for estimating potential credits. Procedures for quantifying credits will obviously be needed.

3) *Expanding the positive list to include qualification of non-carbon dioxide-related project opportunities.* As research into potential greenhouse gas mitigation activities grows, it is becoming increasingly clear that many substantial, cost-effective, and easily quantifiable greenhouse gas reduction opportunities can be undertaken by controlling non-carbon dioxide-related greenhouse gases. Examples include projects that reduce methane and PFC emissions. Thus, the positive list should be expanded to include technologies and processes that focus on greenhouse gases other than carbon dioxide. These should include coal mine methane reduction activities, recovery and

flaring of vented gas, wastewater biogas recovery, landfill gas recovery, and reduction of process emissions.

The Official U.S. Approach

The official U.S. approach provides a much more detailed and comprehensive methodology for evaluating projects than does the EU positive list. Nonetheless, the case study analysis raised several issues that should be addressed to strengthen the U.S. approach. These issues are summarized in the following paragraphs.

Classifying Free Riders: The first and most important issue raised by the case study analysis is the inability of the U.S. approach to effectively screen out business-as-usual projects utilizing conventional technologies. By definition, the "X" percentile threshold test will qualify a certain percentage of conventional technology free riders. The number of free-riders thus qualified will depend on the value of "X." For example, if "X" is set equal to the 20th percentile, roughly 20 percent of new business-as-usual, conventional technology projects will qualify for credits. This is problematic, because the *error* in credits awarded to these free rider projects will equal 100 percent of the awarded credits. Yet, no additional measures have been built into the official U.S. approach to prevent this guaranteed crediting of free riders. The technology matrix approaches do a better job of screening out these free rider projects (see Table 6). Instead of focusing on a "better than average" test, they directly address the additionality of the technologies and processes considered for credits. In particular, the technology matrix approaches make sure that only non-conventional, advanced technologies will qualify for credits under the matrix.

Inclusion of Prime Mover Type: Another additionality classification problem derives from the determination of additionality according to fuel type alone, without factoring in the type of prime mover used in the project. In the power sector, the U.S. additionality test for gasified projects involves a comparison of the plant's emission rate to a threshold that is set using emission rate data from all gas plants, including both steam and gas turbines. As discussed in case studies ES3 and ES5, steam turbines are larger and more efficient than gas turbines. If steam turbines are included in the percentile distribution used for evaluation of a gas turbine project, the average emission rate used for the percentile threshold may be so low that it will exclude most gas turbines from receiving credit. Similarly, an evaluation of a steam turbine project would receive a larger number of credits if gas turbine projects were to be included in the percentile distribution. To avoid excluding truly additional gas turbine projects, such as the gas turbine project in India (ES5), the U.S. approach should be adjusted to allow for an additionality determination that classifies projects according to prime mover type as well as fuel type. The consideration of prime mover type should also apply during the second step of benchmark development.

Non-Energy Efficiency and/or Emission Rate Improvements: The U.S. additionality test is incapable of dealing with projects that cannot be readily classified as either efficiency or emission rate improvements. This issue was illustrated by case study IS3, which involves the installation of variable frequency drives in Argentina. In this case,

29

emission rates and efficiencies were irrelevant because the activity represents an energy conservation project. In this particular case, it would be extremely difficult to establish a percentile threshold in any meaningful way. Perhaps the failure to develop an appropriate additionality test for this type of industrial project stems from the heavy emphasis of the U.S. method on project development in the electricity sector. Once the U.S. approach is developed further, the need for default evaluation tests will become more apparent. For example, the U.S. could establish a back-up test to account for industrial practice project scenarios that fall outside of the "efficiency and/or emissions rate" box.

Similarly, the U.S. additionality encountered problems when applied to residential and commercial sector projects. For several residential and commercial projects it appears meaningless to compare the emission rates of projects because the size, scope, energy usage, and fuel mix of electricity supplied varies greatly between projects. Thus, it may be useful to develop additional criteria for comparing indicators, such as percentage improvement in energy usage or emissions.

Zero-Emission Projects: Finally, the U.S. additionality test, by definition, fails to screen out free rider, zero emission projects. Case study ES8, which involves the construction of a large-scale hydro plant, qualified under the U.S. method although this project is a free rider. The automatic approval of zero emission projects is particularly problematic for the evaluation of large-scale hydro projects and nuclear power plants. These technologies have been developed commercially worldwide, and few if any of them are likely to be undertaken solely to receive emission reduction credits. However, not only does the U.S. method fail to provide an effective test for distinguishing free riders from credit worthy zero-emission projects; it automatically grants credits to all such projects.

Retrofits: The U.S. approach does not clarify how the benchmark should be computed for projects involving retrofits or modifications to existing facilities. Indeed, it is unclear whether the credits for such projects would be derived using a sector or sub-sector benchmark, or the actual emission rate of the affected plant prior to the project (ES2: Heat Rate Improvement Project in India). In the technology matrix approach, a clear distinction is made between new facility projects and projects involving retrofits to existing facilities, and separate credit computation procedures are provided for each of these two classes of projects. Only the former is required to use benchmarks, the latter must utilize the emissions rates of the affected facilities, prior to the retrofit. This approach ensures a more appropriate, and more accurate, emission reduction estimate for retrofit projects. The same distinction between retrofits and new facility projects needs to be made or clarified as part of the U.S. methodology.

Transportation, Commercial, and Residential Sector Projects: The U.S. methodology does not provide guidance on how to evaluate additionality and develop benchmarks in the transportation, residential, and commercial sectors. Our case studies in these sectors represent a preliminary attempt at applying the U.S. approach to this type of projects. In particular, our case studies raise two issues. First, the analyses highlight the need for a definition of what is meant by "recent and comparable" activities in the three sectors. For projects involving replacement of vehicles, such as the natural gas vehicle project in India

(TS1), we interpreted "recent and comparable" to mean all new passenger vehicles sold in 1998. However, by including all new vehicles sold within one year, it becomes almost impossible to devise a percentile test that effectively screens out efficient, but non-additional, vehicles that are already on the market. As witnessed in project TS2 (new gasoline-fueled taxis in India) even a percentile threshold of less than two percent would allow free rider projects to qualify for credits. This suggests that the percentile test may not be the most appropriate additionality screen for this type of projects.

For other types of transportation projects, such as infrastructure or traffic management activities, it would probably be necessary to include a longer time frame for the percentile test, such as three to five years. Thus, it will be necessary to examine the various types of potential projects in the transportation sector and develop guidelines for applying the additionality test to each of these categories.

Moreover, U.S. guidance on estimating credits from retrofit/replacement versus new capacity projects should be clarified. According to the U.S. methodology, it appears that credits would be awarded for a retrofit project in the power sector by subtracting the project emissions from the emissions of the affected power plant. Projects that include new activities or provide new generation would use a sector average. Many potential projects in the transportation sector involve the replacement of old vehicles with new and more efficient models. Likewise, many projects in the residential and commercial sector involve replacement of technologies or retrofits to existing facilities. Similar to what is suggested for projects in the power sector, replacement projects in these sectors should compare their emissions to the emissions of the situation to be replaced, rather than a sector average. Such an analysis will provide a much more accurate estimate. Unlike the situation in the power sector, however, it may be more difficult and costly for project developers to derive accurate emissions data, particularly for projects involving replacement of vehicles. Most developing countries do not collect data on the emissions or fuel economy of vehicles as they age. In these types of situations, project developers would have to obtain specific emissions data for the old vehicles to be replaced, or rely on default emission factors (such as the Intergovernmental Panel on Climate Change (IPCC) data). Guidelines should therefore be developed that detail the types (regional, global, etc.) of default emission factors that can be used.

In addition, the guidelines for developing the baseline for replacement projects should be expanded to distinguish between types of transportation projects. In developing countries, older vehicles rarely employ pollution control technology and have low efficiencies due to poor maintenance. Thus, emissions from older vehicles are generally much higher than emissions from more recent models. It is therefore important to distinguish between projects that are set up to retire and replace old vehicles that have not yet reached the end of their life cycle and projects that intend to substitute the purchase of new conventional vehicles with a more efficient option. The U.S. approach should be expanded to specify that projects involving the retirement and replacement of vehicles that have not yet reached the end of their life cycle should use the project-specific approach for estimating credits, while projects involving the replacement of old vehicles,

31

that would be taken out of service regardless of the prospect for emission reduction credits, should apply a benchmark of recent and comparable activities.

Summary: In summary, there are several steps that can be undertaken to strengthen the official U.S. approach. Recommended steps include:

1) *Clarify the distinction between and treatment of new facility and retrofit projects during benchmark development.* The technology matrix approach makes a clear distinction between new facility projects and projects involving the retrofit of existing facilities. In the latter case, historical data for the affected facility is used to compute the baseline. By using this approach, the credits awarded for retrofit projects will be more accurate. The U.S. approach should utilize the same methodology.

2) *Establish a back-up additionality test that accounts for industrial project scenarios that fall outside the "efficiency and/or emission rate" box.* A number of industrial sector projects that reduce greenhouse gas emissions involve processes that cannot be quantified through efficiency and/or emission rate improvements. Typically, these projects improve the utilization of a specific technology, rather than the efficiency of the usage. As the U.S. methodology is further developed, particular attention should be paid to the development of default thresholds for those types of projects that do not fall into the standard efficiency/emission rate category.

The Technology Matrix

The two technology matrix approaches proved to be slightly more successful methods for classifying the case studies included in this analysis. In total, 22 out of 40 projects were correctly identified as either additional or free riders using the technology matrix. Moreover, the technology matrix is more successful at providing a definite determination of the classification status of the projects examined. The technology matrix resulted in less than half as many indeterminate or unclear classifications than did the U.S. approach.

In addition, the technology matrix will probably less costly to implement than the U.S. methodology, given that the latter method has more extensive data requirements. As discussed above, the U.S. methodology relies heavily on sector-wide data for both the additionality test and benchmark development. Thus, if data is unavailable or non-existent, it will be impossible to evaluate the additionality of projects by using the U.S. approach. In contrast, the technology matrix is much less dependent on data availability for qualifying technologies, thereby lowering the financial requirements for implementing the CDM, or a similar flexible mechanism approach.

In spite of these advantages, the case studies revealed a number of issues regarding the technology matrix that still need to be addressed before this methodology can be implemented properly. The first issue relates to the process of evaluating the additionality of technologies to be included in the matrix. The two major criteria for evaluating additionality include an assessment of the economics and market penetration of the technology in question. The case study ES4, involving the creation of a natural gas

32

combined cycle (NGCC) plant in India, raised an important issue regarding the market penetration test. This test is used to determine whether certain barriers, such as lack of investment, knowledge or technical capacity, have previously prevented a technology from being applied in a specific country. The assumption is that such barriers can be identified by first determining whether the technology has been unable to gain market access. If it is determined that the technology is not currently available in a given country, it will likely mean that it is additional. In this way, the technology matrix's market penetration test will always qualify first-of-its-kind projects, although the technology applied is commercial and widely available in other countries. This was exactly the result achieved for project ES4 where the technology matrix approaches qualified the NGCC plant for credit, despite its status as a free rider.

The case study analysis revealed another problem with the technology matrix approach. The technology matrix is set up to evaluate *all* of the processes included in a project as a single entity. Consequently, the approach is ill equipped to deal with a project, such as the Highly Efficient Fertilizer Complex in Jordan (IS9), which involves multiple processes. Since, for IS9, only one of the processes involves the utilization of advanced technologies, it is unclear whether the project as a whole should be accepted or rejected for credits. The same problem arose in the first commercial project, involving energy efficiency and conservation measures in commercial buildings in the Philippines (CS1), where only one of the project components involved an advanced technology.

Several steps can be undertaken to strengthen the technology matrix approach. Recommended steps include:

1. *Strengthen market penetration test to effectively evaluate first-of-its-kind projects.* A concern with the additionality test of the technology matrix is that the market penetration test will tend to qualify first-of-its-kind projects, regardless of whether such projects are additional or free riders. The answer may be to use a global or regional market penetration test. However, one drawback of this approach would be that the improved stringency of the technology matrix would disqualify technologies that are truly additional in some countries. Still, the feasibility of applying a regional or world wide market penetration test should be examined to determine what effect it may have on the ability of the technology matrix to screen out free riders.

2. *Reexamine the technology matrix to accommodate projects involving installation of advanced equipment in only one process.* Currently, the technology matrix has been designed to evaluate an entire project at once. Thus, projects that deploy advanced technologies in only one process out of many cannot be evaluated properly under the matrix approach. One solution would be to account for the emission reductions from the advanced non-commercial technology and qualify just that part of the project. However, procedures for undertaking such an analysis would have to be specified, and guidelines for establishing the benchmark should be developed.

Land Use and Forestry Sector Projects

Land use and forestry sector projects are inadequately addressed by all four of the baseline methodologies. The U.S. methodology provides only minimal guidance for determining additionality for land use projects. The U.S. approach indicates that for land use projects involving carbon sinks, the "eligibility threshold would represent activities that are better than the prevailing conditions within a country or region." Further, it states that "since natural variability may cause sequestration areas to vary immensely, the threshold of performance may require demonstrating divergence from a regional trend."[8] However, this methodology is problematic because it is unclear what is meant by "activities," "prevailing conditions," and a "regional trend." In particular, it is unclear whether the project should be compared to a threshold derived from similar activity types or similar land use/forestry plots. Moreover, if similar forest activities are not being conducted in the area, it is uncertain whether it is necessary to find data on similar activities from another country or if it is sufficient to compare the project activities to current sequestration on other land areas (even if these areas are not undergoing similar forest management activities) within the region. Clearly, specified directions for when to use similar activity types versus similar land plots should be developed, including default procedures for which activities to include if there are no similar activities to use for the comparison.

In addition, if the U.S. intends its eligibility threshold test to be applied to the land use sector, then it must clearly indicate the necessary data for use in establishing the threshold. As currently written the U.S. approach does not provide guidance on the necessary data that should be used for developing an eligibility threshold. For the industrial sector, for example, the U.S. approach clearly indicates that the eligibility threshold should be set at "X" percentile of efficiencies or emissions rates. There is a lack of comparable language for land use projects. It is not even clear whether land use/forestry sector projects should be analyzed based on a comparison of the rate of carbon sequestered (expressed as annual carbon sequestration of the project or annual carbon sequestration per hectare) or a comparison based on a more subjective analysis, such as an analysis of trends in project types.

The process of estimating credits under the U.S. approach is less problematic than undertaking the additionality test. In fact, the U.S. approach is clear in its statement that "baselines are based on the current situation" when calculating credits.[9] Thus, typically, the total amount of CO_2 sequestered in absence of the project is subtracted from the total amount of CO_2 to be sequestered with the project.

---

[8] SAIC, " *Political Analysis of the U.S. and EU Market Mechanism Proposals: Subtask 1 Final Working Paper*," December 2000, pg. 11.

[9] SAIC, " *Political Analysis of the U.S. and EU Market Mechanism Proposals: Subtask 1 Final Working Paper*," December 2000, pg. 11.

A lack of guidance also exists under the technology matrix approach for determining additionality and awarding credits for projects falling within the land use sector. As it currently stands, the technology matrix approach does not provide any mention of specific additionality criteria for land use projects. Most land use and forestry projects do not include advanced technologies. Thus, this type of project has not been included in previous work on developing the technology matrix. It is very possible that a land use project may include use of an advanced, non-commercial "process;" however, it is unclear whether the technology matrix intends for advanced, non-commercial "processes," as well as technologies to qualify as additional. Indeterminate qualifications of additionality may be avoided if the technology matrix is revised to clarify whether advanced, non-commercial "processes" in the land use sector may be considered along with advanced, non-commercial technologies. Because it lacks clear guidance on both additionality and baseline determinations, the technology matrix should be updated to specify that land use and forestry projects involving advanced technologies or processes may use the technology matrix approach for the evaluation of additionality, and the project-specific approach for estimation of credits. It should also specify that all other projects that are not applicable to the market penetration and economic feasibility tests under the technology matrix should use the project-specific approach.

It is important to emphasize that no matter what approach is used in regards to land use projects, carbon sequestration data can vary greatly within a country or region. Often, comparisons of lands of similar area and even tree/plant species yield entirely divergent carbon sequestration totals (see project LU2, Afforestation of Marginal Agricultural Land in Russia). It has been argued that accounting for changes in carbon stocks in land use projects is inherently more difficult than accounting for carbon emissions in the power sector. Two significant problems are resolution (recognizing small changes in large numbers) and maintaining the infrastructure needed for regular measurement of changes in carbon stocks. Temporal and spatial variability cause high variability in soil carbon estimates at all scales.[10] Simply put, forest carbon stocks are incredibly varied, depending on latitude, climate, ecosystem (i.e. tropical, temperate, boreal), species mix, and soil regime.[11]

As noted in LU1 (Forest Protection and Management in Six Mexican Communities), other issues are inherent to land use sector projects that may undermine the accuracy of additionality and baseline determinations; most notably, "leakage" and "permanence." It is important to recognize that, no matter which approach is used, land use projects possess many unique characteristics, such as "leakage" and "permanence," which may undermine the accuracy of additionality and baseline determinations. Leakage is defined as "the unexpected loss of anticipated carbon benefits resulting from additional effects of the project's activities outside the project boundaries." For example, a project designed to prevent deforestation may result in persons moving elsewhere and deforesting other land,

---

[10] Pew Center on Global Climate Change, "Land Use and Global Climate Change: Forests, Land Management, and the Kyoto Protocol," June 2000, p. 11.
[11] WRI "*Getting It Right: Emerging Markets for Storing Carbon in Forests*," 1999.

resulting in little to no additional carbon savings (i.e., activity shifting). Further, a project designed to reduce forest harvesting may result in the increase of forest harvesting in another region to satisfy demand (i.e., market effects). Leakage is typically associated with a loss in carbon, but in some instances, leakage can be positive when projects lead to more carbon benefits than initially estimated.

Permanence is defined as "the possibility of a reversal of carbon benefits from either natural disturbances such as fires, disease, pests, and unusual weather events; or from the lack of reliable guarantees that the original land use activities will not return." Factors such as drought, frost, weeds, foraging animals, insects, infestation, wind and water erosion, fire, and other unanticipated anthropogenic disturbances could damage afforested or reforested land, for example, and cause carbon sequestration to be lost or reversed in future years. As stated, these uncertainties make it difficult to ensure accuracy in additionality and baseline determinations and to develop a standardize methodology. Although a project may be correctly identified as additional, the benefits of the project may, for example, be reversed by individuals increasing harvesting or deforesting elsewhere or lost by an unforeseen natural disturbance, such as fire. Therefore, a positive additionality determination or the award of credits may later prove erroneous if the estimated carbon sequestration is actually reversed through the problem of permanence or lost through the problem of leakage.

Because of the high degree of uncertainty involved in accounting for carbon sequestration in land use and forestry sector projects, these projects do not lend themselves easily to standardized baseline methodologies. For this type of project, the project-specific approach provides a better alternative. This approach does not rely upon regional averages, which vary greatly, but instead focuses on what would have occurred in absence of the project. Use of this approach is likely to yield a more accurate baseline determination.

In summary, several steps can be undertaken to strengthen the official U.S. and technology matrix approaches in regards to land use and forestry sector projects. Recommended steps include:

1) *The U.S. approach should be revised to clarify and fully define terminology and concepts within its approach to determining additionality of land use projects.* Specifically, the U.S. approach should clarify what is meant by "activities," "prevailing conditions," and a "regional trend." Clearly specified directions for when to use similar activity types versus similar land plots should be developed, including default procedures for which activities to include if there are no similar activities to use for the comparison. The U.S. approach should clearly indicate which type of data is needed for use in establishing a percentile threshold.

2) *The technology matrix approach should be revised to include an approach for making additionality determinations and awarding credits to land use sector projects.* Specifically, the technology matrix approach should clarify whether advanced, non-commercial "processes" in the land use sector may be considered equal to advanced,

non-commercial technologies in the additionality determination. The approach should be updated to specify that land use and forestry projects involving advanced technologies or processes may use the technology matrix approach for the evaluation of additionality, and the project-specific approach for the estimation of credits. All other projects that are not applicable to the technology matrix market penetration and economic feasibility tests should use the project-specific approach.

Conclusion

All of the standardized methodologies analyzed encountered problems at various points during this case study analysis. Two issues are particularly relevant for all four methodologies. First, each methodology will require a back-up methodology for situations where standardized approaches cannot be applied. The full technology matrix specifically recommends using the project-specific approach as the fallback. The other baseline approaches should specify equivalent fallback procedures. Second, every standardized baseline approach will be affected by the problem of deficient and missing data in host countries. However, the official U.S. approach will be particularly affected, due to the data requirements of developing a percentile threshold test based on a distribution of "recent and comparable" data.

Many of the other issues raised were more specific to each of the four baseline methodologies. Several steps have been outlined above to strengthen each project evaluation approach. In general, the technology matrix approaches encountered the least difficulties. The technology matrix thus far appears to offer a slightly improved technical solution to the many challenges of project evaluation.

The ultimate goal of any future program, be it voluntary or mandatory, domestic or international, will be to reduce overall global greenhouse gas emissions. The implementation of a project evaluation approach that minimizes additionality classification errors is crucial to this goal. The case study analysis indicates that the technology matrix offers several advantages. First, it specifically provides for the incorporation of an alternative methodology for project situations where the matrix does not apply or is unable to provide an accurate emission reduction estimate. However, the combination of the technology matrix with the project-specific approach offers one significant drawback. When a project is small, the transaction costs may be too high to warrant use of the project-specific approach. Thus, some truly additional projects may be disqualified. A second advantage of the technology matrix approach is that it requires less data for undertaking the additionality evaluation and baseline development than does the official U.S. approach. Finally, the technology matrix is technology neutral in the sense that it focuses on the environmental additionality of the activities examined rather than relying on political processes to determine the emissions threshold or an acceptable technology.

APPENDIX A

PROJECT NUMBER: ES1

COUNTRY: India

SECTOR: Electricity

PROJECT TITLE: IGCC Power Plant

PROJECT DESCRIPTION: This project involves the construction of a new, 500-MW IGCC power plant in India. The power plant is needed to meet India's rapidly growing demand for electricity. The plant will utilize coal as its primary fuel, and will operate as a baseload facility. The plant will be built as a joint venture between an Indian utility and an U.S. investor-owned utility. The U.S. utility will receive all of the credits to be awarded to the project, along with a share of the project's ownership, in exchange for its financial backing.

PROJECT ADDITIONALITY: Currently, coal-fired IGCC for power generation is an advanced combustion technology that is not being used on a commercial basis, either in India or elsewhere (Note, however, that oil-fired IGCC is being utilized, particularly for applications at petroleum refineries). The project developers decided to use IGCC rather than conventional technology in order to obtain the credits that would be available to an advanced-technology project under an international carbon offset program. Therefore, the project is additional.

PROJECT EMISSIONS: The power plant is expected to operate with an average heat rate of 7560 Btus/kWh. Based on an emission factor of 205.3 lbs CO_2/mmBtu for bituminous coal, the project's emission rate (ER) is estimated as follows:

ER = (7560 Btus/kWh)(205.3 lbs CO_2/mmBtu)(1 mmBtu/1,000,000 Btus)

ER = 1.55 lbs CO_2/kWh

PROJECT BENCHMARK DATA: Seven coal-fired power plants have been opened in India since 1995. The EPA has collected heat rate data on these and older power plants. Two of the post-1995 power plants have heat rate data that are suspect: 7365 and 5611 Btus/kWh. These two power plants were therefore eliminated as outliers. The heat rate data for the remaining 5 power plants are as follows:

Utility	Power Plant	Capacity (MW)	Year Commissioned	Heat Rate* (Btus/kWh)	Emissions Rate* (lbs CO_2/kWh)
NTPC	Talcher	1000	1995	10,015	2.06
Orissa	1b Valley	420	1995	10,218	2.10
Damodar Valley	Mejia	630	1995-96	9,745	2.00
BSES Ltd.	Dahanu	500	1995	10,610	2.18
Bihar SEB	Tenughat	420	1996	10,466	2.15
Total/Average	NA	2970	NA	10,211 (10,150)	2.10 (2.09)

*Averages shown are arithmetic. Weighted averages are in parentheses.

PROJECT ANALYSIS TABLE: Project Number ES1

	U.S. Proposal	EU Proposal	Full Technology Matrix	Hybrid Technology Matrix
Does project qualify?	If "X" > 20th percentile, then the threshold will be at least 2.00, and the project will qualify (ER<2.00). If X < 20th percentile then threshold test must be based on regional rather than Indian data, but it is likely that project will qualify as additional regardless of data used (because IGCC emissions rate much less than conventional coal emissions rate).	Clean coal projects will not qualify under the positive list unless they have efficiencies > 55 percent. This project's efficiency is only 45 percent, so it will not qualify for credits	IGCC is an advanced, non-commercial technology. Projects using this technology will automatically qualify as additional under the technology matrix.	Project will automatically qualify as additional.
Is the project correctly identified as either a free rider or an additional project?	Yes	No. Project is a coal project, and such projects are not included under the EU's positive list.	Yes	Yes
Number of credits Awarded	The benchmark would be taken as either the weighted average of the five Indian plants (2.09) or the average of a larger set of regional plants. In the former case, the estimated credits would be 2.09-1.55 = 0.54 lbs/kWh	Not applicable.	Estimated credits = 2.10 – 1.55 = 0.55 lbs/kWh	2.09 – 1.55 = 0.54 lbs/kWh
Error in credits Awarded	Unknown.	Not applicable.	Unknown.	Unknown.

METHODOLOGY ASSESSMENT: This additional project will qualify for credits under all approaches except the EU approach. Under the EU positive list of technologies, the project will fail to qualify not because it is deemed non-additional, but because it is a coal project. The goal under the EU approach is to push developing countries away from coal and other fossil fuels, and towards renewables. Note, however, that it is unlikely that the project developers would opt for renewables in lieu of IGCC. The power plant is being built to meet a specific identified market need—i.e., a need for a large capacity (500-MW) baseload plant to serve rapidly growing demand. Renewable technologies such as solar and wind cannot be used for such large-capacity, baseload applications. If disqualified under the EU positive list test, it is likely that the project developers would build a conventional coal-fired plant in place of the IGCC plant. Rarely would disqualification under the positive list cause project developers to opt for a renewables plant in lieu of a fossil fuel plant, because renewables technology, even if feasible at or near the project site, serve a different market application. In short, the EU's goal of changing clean development paths via the positive list is not likely to prove successful.

Under the full technology matrix approach, the project will be awarded credits at the rate of 0.55 lbs per kWh generated. This project reduction estimate is based on a benchmark reflecting the average emission rate of all recently built coal-fired power plants in India. The official U.S. approach would utilize a benchmark reflecting the weighted average emission rate, and will award credits at the rate of 0.54 lbs per kWh generated, but *only* if "X" in the percentile threshold test were set equal to or greater than the 20th percentile. If X < 20th percentile, it would be necessary to define both the threshold and the benchmark on the basis of regional data (perhaps, e.g., including data for China) rather than data specific to India. This example illustrates the fact that data requirements are more difficult to meet under the official U.S. approach than under the technology matrix approach. Under the former approach, data sufficient to define a percentile distribution are required.

PROJECT NUMBER: ES2

COUNTRY: India

SECTOR: Electricity

PROJECT TITLE: Heat Rate Improvement

PROJECT DESCRIPTION: In India, electricity demand exceeds supply by 10 to 20 percent, and demand is growing rapidly. Due to lack of capital, India cannot build enough new power plants to close the supply-demand gap. In part for this reason, the Government of India's current Five-Year Plan calls for the refurbishment of existing power plants, as a less expensive alternative to the installation of new capacity. This project is being undertaken as part of the Five-Year Plan. It involves a heat rate improvement project at an existing, 270-MW conventional coal-fired power plant. The plant is owned and operated by an Indian utility. The goal of the project is to restore the plant to its original, design efficiency. The resulting efficiency improvement will enable the power plant to generate more electricity for sale to the grid. The project will involve the replacement of many existing, worn-out plant components (turbine blades, condenser tubing, etc.) with new equipment. However, the project will utilize existing, commercial technologies exclusively.

The project is being financed primarily by the Indian utility. However, a limited amount of additional financing is being provided by an U.S. investor-owned utility, in exchange for the credits to be awarded to the project under an international carbon offset program.

PROJECT ADDITIONALITY: The project is being undertaken as part of India's Five-Year Plan. The U.S. company's financial assistance is limited and, though welcome by the Indian utility, it is not necessary to the completion of the project. The project is therefore not additional.

PROJECT EMISSIONS: The power plant's current heat rate is 10,000 Btus/kWh. The project is expected to reduce this heat rate by 3.5 percent, to 9650 Btus/kWh. Based on an emission factor of 205.3 lbs CO_2/mmBtu for bituminous coal, the project's emission rate (ER) is estimated as follows:

$$ER = (9650 \text{ Btus/kWh})(205.3 \text{ lbs } CO_2/\text{mmBtu})(1 \text{ mmBtu}/1,000,000 \text{ Btus})$$

$$ER = 1.98 \text{ lbs } CO_2/\text{kWh}$$

PROJECT BENCHMARK DATA: A total of seven coal-fired power plants have been opened in India since 1995. The EPA has collected heat rate data on these and older power plants. Two of the post-1995 power plants have heat rate data that are suspect: 7365 and 5611 Btus/kWh. These two power plants were therefore eliminated as outliers. The heat rate data for the remaining 5 power plants are as follows:

42

Utility	Power Plant	Capacity (MW)	Year Commissioned	Heat Rate* (Btus/kWh)	Emissions Rate* (lbs CO_2/kWh)
NTPC	Talcher	1000	1995	10,015	2.06
Orissa	1b Valley	420	1995	10,218	2.10
Damodar Valley	Mejia	630	1995-96	9,745	2.00
BSES Ltd.	Dahanu	500	1995	10,610	2.18
Bihar SEB	Tenughat	420	1996	10,466	2.15
Total/Average	NA	2970	NA	10,211 (10,150)	2.10 (2.09)

*Averages shown are arithmetic. Weighted averages are in parentheses.

PROJECT ANALYSIS TABLE: Project Number ES2

	U.S. Proposal	EU Proposal	Full Technology Matrix	Hybrid Technology Matrix
Does project qualify?	If "X" > 20th percentile, then the threshold will be at least 2.00, and the project will qualify (ER<2.00). If X < 20th percentile then threshold test must be based on regional rather than Indian data.	Under the EU's positive list, projects to rehabilitate fossil fuel plants must introduce a technology change that improves efficiency by at least 5 percent. This project does not involve a technology change, and the resulting efficiency improvement is less than 5 percent. Therefore, the project does not qualify as additional.	The project does not involve non-commercial technology, and will not qualify under the technology matrix. The project developers may prove additionality using the project-specific approach. However, given that the project is part of a pre-existing plan, it is unlikely that additionality could be proved.	The project will not qualify as additional under the technology matrix, and is unlikely to qualify as additional under the project-specific approach.
Is the project correctly identified as either a free rider or an additional project?	No if X >20th percentile; Possibly if X < 20th percentile. The number of free riders qualified under the U.S. approach depends on the value that is chosen for "X."	Yes	Yes	Yes
Number of credits Awarded	It is unclear whether credits for a heat rate improvement project would be computed using a benchmark or the actual heat rate of the plant prior to the project. If the former, then credits = 2.09-1.98 = 0.11 lbs CO_2/kWh. If the latter, credits = (10,000-9650)(205.3) /1,000,000 = 0.07 lbs CO_2/kWh	Project does not qualify for any credits	Project does not qualify for any credits	Project does not qualify for any credits
Error in credits Awarded	The project is a free rider; thus the error is equal to 100 percent of the credits awarded (either 0.07 or 0.11 lbs/kWh)	The project is correctly identified as a free rider; therefore the error in the credits awarded is zero.	The project is correctly identified as a free rider; therefore the error in the credits awarded is zero.	The project is correctly identified as a free rider; therefore the error in the credits awarded is zero.

44

METHODOLOGY ASSESSMENT: This project is correctly identified as a free rider by all of the methods except the official U.S. method. It is an example of a "conventional-technology free rider." By its very nature, the X percentile threshold test will qualify a certain percentage of business-as-usual projects utilizing conventional technologies. The number of conventional-technology free riders thus qualified will depend on the value of X. For example, if X is set equal to the 20th percentile, we can expect that roughly 20 percent of new business-as-usual, conventional-technology projects will qualify under the carbon offset program. For the power generation sector, both the EU's positive list and the technology matrix do a better job of screening out conventional technology free riders.

Not only does the official U.S. method fail to screen out this free rider project, but also further errors will be generated by this method during baseline development. It is unclear, based on the existing documentation, whether or not the baseline for heat rate projects is to be computed using the actual heat rate of the plant prior to project initiation, or a sector benchmark. As this project demonstrates, the use of a sector benchmark would nearly double the size of the error in the credits awarded, from 0.07 lbs CO_2/kWh, to 0.11 lbs CO_2/kWh.

The technology matrix approach makes a clear distinction between new facility projects and projects involving retrofits to existing facilities. Only the former projects may utilize sector benchmarks; the emission baseline for a retrofit project must be computed using historical data for the affected facility. The official U.S. method should also be either modified, or better explained, to make or clarify this same fundamental distinction between new facility and retrofit projects.

PROJECT NUMBER: ES3

COUNTRY: India

SECTOR: Electricity

PROJECT TITLE: Fuel Switching

PROJECT DESCRIPTION: India's power sector is plagued by chronic coal supply and delivery problems. Due to inadequate mine capacity, transportation bottlenecks, port capacity constraints, and other problems, coal-fired plants must often be shut down due to a lack of fuel. The resulting loss of generation contributes to India's chronic electricity supply shortfall and the consequent rolling blackouts.

Let us suppose that a new natural gas pipeline has been built from Uzbekistan, through Afghanistan and Pakistan, to western India. Consequently, it is now possible to convert an existing coal-fired power plant, located near the pipeline, to dual-firing capability. The converted plant will continue to use domestic Indian coal when it is available, because it is cheaper than the imported gas. However, the plant will switch to gas in order to keep operating during coal supply emergencies.

The project is treated as a top priority by the Indian government, because it is expected to reduce the frequency and duration of rolling blackouts by approximately 50 percent in the plant's service territory. However, although the government is committed to fully financing the project with or without foreign aid, an opportunity for attracting additional investment to the project is recognized. By promising all credits generated by the project to foreign investors, the government obtains additional project financing from an U.S. investor-owned utility.

PROJECT ADDITIONALITY: The project is being undertaken to reduce electricity supply shortfalls, not to reduce emissions or gain credits. The project would be undertaken with or without the U.S. company's involvement; hence, it is a non-additional free rider.

PROJECT EMISSIONS: The power plant's heat rate is 10,000 Btus/kWh. Hence using the natural gas emissions factor of 117.1 lbs CO_2/mmBtu, the emission rate for the project can be computed as follows:

$$ER = (10,000 \text{ Btus/kWh})(117.1 \text{ lbs } CO_2/\text{mmBtu})(1 \text{ mmBtu}/1,000,000 \text{ Btus})$$

$$ER = 1.17 \text{ lbs } CO_2 \text{ reduced for each kWh generated using natural gas}$$

PROJECT BENCHMARK DATA: Data on gas-fired power plants in India are unavailable. Therefore, in the following table we have created a number of fictional gas-fired power plants, and added them to the real-world coal-fired plants first introduced in Project Numbers ES1 and ES2.

Utility	Power Plant	Prime Mover*	Fuel Type	Capacity (MW)	Year Commissioned	Heat Rate** (Btus/kWh)	Emissions Rate** (lbs CO_2/kWh)
NTPC	Talcher	ST	Coal	1000	1995	10,015	2.06
Orissa	1b Valley	ST	Coal	420	1995	10,218	2.10
Damodar Valley	Mejia	ST	Coal	630	1995-96	9,745	2.00
BSES Ltd.	Dahanu	ST	Coal	500	1995	10,610	2.18
Bihar SEB	Tenughat	ST	Coal	420	1996	10,466	2.15
Total/Average	**NA**	**ST**	**Coal**	**2970**	**NA**	**10,211 (10,150)**	**2.10 (2.09)**
ACME	Gas 1	ST	Gas	400	1995	10,500	1.23
ACME	Gas 2	ST	Gas	500	1997	10,100	1.18
ACME	Gas 3	ST	Gas	1000	1999	9,500	1.11
Total/Average	**NA**	**ST**	**Gas**	**1900**	**NA**	**10,033 (9,868)**	**1.17 (1.15)**
Small Op	Plant A	GT	Gas	10	1995	16,000	1.87
Small Op	Plant B	GT	Gas	15	1995	16,100	1.89
Small Op	Plant C	GT	Gas	5	1995	15,700	1.84
Small Op	Plant D	GT	Gas	5	1996	14,500	1.70
Small Op	Plant E	GT	Gas	10	1997	15,600	1.83
Small Op	Plant F	GT	Gas	10	1997	16,200	1.90
Small Op	Plant G	GT	Gas	10	1997	15,300	1.79
Total/Average	**NA**	**GT**	**Gas**	**65**	**NA**	**15,629 (15,746)**	**1.83 (1.85)**
Gas Total/Average	**NA**	**All**	**Gas**	**1965**	**NA**	**13,950 (10,063)**	**1.63 (1.18)**
Fossil Total/Average	**NA**	**All**	**All**	**4935**	**NA**	**12,704 (9,270)**	**1.79 (1.58)**

*ST = Steam Turbine. GT = Gas Turbine. **Averages shown are arithmetic. Weighted averages are in parentheses.

PROJECT ANALYSIS TABLE: Project Number ES3

	U.S. Proposal	EU Proposal	Full Technology Matrix	Hybrid Technology Matrix
Does project qualify?	If X > the 20th percentile, then this project will qualify (ER<1.18). If the 10th percentile < X < the 20th percentile, then the project will not qualify. If X < the 10th percentile, then there is not enough data to determine project eligibility under the U.S. proposal.	This project will not qualify under the EU's positive list, as it does not fall under the categorical listings for renewables, energy efficiency, or demand side management.	This project does not employ advanced, non-commercial technology, so it will not qualify under the full technology matrix. The project developer would be given the opportunity to prove additionality using the project-specific approach, but given that the project is a free rider, it is unlikely that additionality could be proved.	This project will not qualify under the hybrid technology matrix.
Is the project correctly identified as either a free rider or an additional project?	Yes, if the 10th percentile < X < the 20th percentile. No, if X > the 20th percentile.	Yes	Yes	Yes
Number of credits Awarded	If X > the 20th percentile, then the number of credits awarded would be determined by subtracting the project emissions from the plant's emissions rate prior to the fuel switch.* Thus, the credits awarded = 2.05-1.17=0.88 lbs CO_2/kWh.	Not applicable.	Because this project does not qualify under the technology matrix, it does not qualify for credits (i.e., 0 credits would be awarded).	Because this project does not qualify under the hybrid technology matrix, it does not qualify for credits (i.e., 0 credits would be awarded).
Error in credits Awarded	If X > 20th percentile, then the error in credits awarded is equal to 100 percent of the credits awarded (0.88 lbs CO_2/kWh).	Not applicable.	Project is a free rider, so the error in credits awarded is 0.	Project is a free rider, so the error in credits awarded is 0.

*ER prior to fuel switch = (10,000 Btus/kWh)(205.3 lbs CO_2/mmBtu)(1 mmBtu/1,000,000 Btus) = 2.05 lbs CO_2/kWh. Fuel emissions factor of 205.3 lbs CO_2/mmBtu was derived from Energy Information Administration (EIA), *Instructions for Form 1605*, February 2000, pgs. 47-48.

METHODOLOGY ASSESSMENT: This non-additional, free rider project fails to qualify for credits under all approaches except for certain instances under the U.S. approach. This is a replacement project, so credits are awarded by subtracting the project emissions from the emissions of the coal plant that the retrofit project would be replacing.

While the EU and technology matrix approaches each succeed in blocking this non-additional, free rider project from qualifying for credits, the U.S. approach clearly demonstrates a weakness. The plant in question is a coal-fired steam turbine plant. The U.S. additionality test involves the comparison of the plant's emissions rate to a threshold that is set using emissions rate data from all gas plants, including both steam and gas turbines. The determination of additionality according to fuel type only versus prime mover type is a major flaw in the U.S. approach. Steam turbines and gas turbines are markedly different in nature. Gas turbine plants are peak load plants, meaning that they are only in operation when there is a demand for electricity. These plants are smaller, more abundant, and less efficient than steam turbine plants. Steam turbine plants, on the other hand, are base load plants, meaning that they are designed to operate at all times. Steam turbine plants are larger and more efficient than gas turbine plants.

In this example, the emissions rates for the three steam turbine plants for which data was available are notably lower than for the gas turbine plants. The inclusion of the gas turbine data points in this determination acts to weight the distribution to be more comparable to a gas turbine plant. If the threshold were set using only the steam turbine data, then this project would likely not have qualified. The U.S. should perhaps consider adjusting the additionality test to allow for a more comparable test of plants, providing for the comparison of prime mover type as well as fuel type.

COUNTRY: India

SECTOR: Electricity

PROJECT TITLE: Natural Gas Combined Cycle

PROJECT DESCRIPTION: The Indian government has been encouraging the construction of natural gas-fired power plants in recent years, due in part to the capital cost advantages of gas-fired technologies. However, India's limited domestic gas supplies has constrained its ability to take full advantage of these technologies.

Let us suppose that the new natural gas pipeline, introduced in Project Number ES3, makes possible the building of gas-fired power plants in some parts of western India. This project involves the construction of one such plant: a 500-MW natural gas combined cycle facility. At present, there are no combined cycle power plants in India, although some are planned. The Indian utility has decided to make this project its first foray into combined cycle technology. An U.S. investor-owned utility has agreed to provide 5 percent of the financing required for the project, in exchange for the project's credits.

PROJECT ADDITIONALITY: The Indian utility was faced with a choice of building either a steam turbine plant or a combined cycle plant. Although the foreign financing acted as an added incentive for choosing the latter, the decision was based primarily on the proven capital and operating cost advantages of combined cycle technology. These advantages were judged to be worth the risk of introducing the technology into India for the first time. Since the incentives provided by an international carbon offset program were not the primary factor in the decision to use combined cycle technology, the project is not additional.

PROJECT EMISSIONS: The power plant's heat rate is expected to be 7500 Btus/kWh. Hence using the natural gas emissions factor of 117.1 lbs CO_2/mmBtu, the emission rate for the project can be computed as follows:

$$ER = (7500 \text{ Btus/kWh})(117.1 \text{ lbs } CO_2/\text{mmBtu})(1 \text{ mmBtu}/1{,}000{,}000 \text{ Btus})$$

$$ER = 0.88 \text{ lbs } CO_2/\text{kWh}$$

PROJECT BENCHMARK DATA: Data on gas-fired power plants in India are unavailable. Therefore, in the following table we have created a number of fictional gas-fired power plants, and added them to the real-world coal-fired plants first introduced in Project Numbers ES1 and ES2.

Utility	Power Plant	Prime Mover*	Fuel Type	Capacity (MW)	Year Commissioned	Heat Rate** (Btus/kWh)	Emissions Rate** (lbs CO_2/kWh)
NTPC	Talcher	ST	Coal	1000	1995	10,015	2.06
Orissa	1b Valley	ST	Coal	420	1995	10,218	2.10
Damodar Valley	Mejia	ST	Coal	630	1995-96	9,745	2.00
BSES Ltd.	Dahanu	ST	Coal	500	1995	10,610	2.18
Bihar SEB	Tenughat	ST	Coal	420	1996	10,466	2.15
Total/Average	NA	**ST**	**Coal**	**2970**	NA	**10,211** **(10,150)**	**2.10** **(2.09)**
ACME	Gas 1	ST	Gas	400	1995	10,500	1.23
ACME	Gas 2	**ST**	Gas	500	1997	10,100	1.18
ACME	Gas 3	ST	Gas	1000	1999	9,500	1.11
Total/Average	NA	**ST**	**Gas**	**1900**	NA	**10,033** **(9,868)**	**1.17** **(1.15)**
Small Op	Plant A	GT	Gas	10	1995	16,000	1.87
Small Op	Plant B	GT	Gas	15	1995	16,100	1.89
Small Op	Plant C	GT	Gas	5	1995	15,700	1.84
Small Op	Plant D	GT	Gas	5	1996	14,500	1.70
Small Op	Plant E	GT	Gas	10	1997	15,600	1.83
Small Op	Plant F	GT	Gas	10	1997	16,200	1.90
Small Op	Plant G	GT	Gas	10	1997	15,300	1.79
Total/Average	NA	**GT**	**Gas**	**65**	NA	**15,629** **(15,746)**	**1.83** **(1.85)**
Gas Total/Average	NA	All	Gas	1965	NA	13,950 **(10,063)**	1.63 **(1.18)**
Fossil Total/Average	NA	All	All	4935	NA	12,704 **(9,270)**	1.79 **(1.58)**

*ST = Steam Turbine. GT = Gas Turbine. **Averages shown are arithmetic. Weighted averages are in parentheses.

PROJECT ANALYSIS TABLE: Project Number ES4

	U.S. Proposal	EU Proposal	Full Technology Matrix	Hybrid Technology Matrix
Does project qualify?	If $X > 10^{th}$ percentile, then the threshold will be at least 1.11, and the project will qualify (ER<1.11). If $X < 10^{th}$ percentile, then there is not enough data to determine project eligibility under the U.S. proposal.	This project qualifies as additional under EU's positive list. NGCC falls under the EU's energy efficiency category, as natural gas combined cycle is an advanced technology for gas fired power plants.	This project qualifies as additional under the Full Technology Matrix, as natural gas combined cycle is an advanced, non-commercial technology in India.	Project will automatically qualify as additional.
Is the project correctly identified as either a free rider or an additional project?	No, if $X > 10^{th}$ percentile. Possibly, if $X < 10^{th}$ percentile. The number of free riders qualified under the U.S. approach depends on the value that is chosen for "X."	No. EU approach does not analyze factors other than whether the project employs renewables or demand side management, or enhances energy efficiency in the qualification determination.	No. The market penetration test under the technology matrix will always qualify first-of-its-kind projects, such as this one, as additional.	No. The market penetration test under the technology matrix will always qualify first-of-its-kind projects, such as this one, as additional.
Number of credits Awarded	The number of credits awarded would be determined by subtracting the project ER from the weighted fossil average (credits=1.58-0.88 = 0.7 lbs CO_2/kWh).	Not applicable.	Estimated credits are derived by subtracting the ER of the project (0.88) from the average ER of the alternative technology. For NGCC in India, it has previously been determined that the most likely alternative technology would be gas-fired steam turbine.* (average ER=1.17) Thus, the number of credits awarded would be: 1.17-0.88=0.29 lbs CO_2/kWh.	Estimated credits for this project are derived by subtracting the project ER from the fossil average of recent coal and gas facilities (1.58). Credits=1.58-0.88 = 0.7 lbs CO_2/kWh.
Error in credits Awarded	The project is a free rider, thus the error in credits awarded is equal to 100% of the credits awarded (i.e., 0.7 lbs CO_2/kWh).	Not applicable.	The project is a free rider, thus the error in credits awarded is equal to 100% of the credits awarded (i.e., 0.29 lbs CO_2/kWh).	The project is a free rider, thus the error in credits awarded is equal to 100% of the credits awarded (i.e., 0.7 lbs CO_2/kWh).

*SAIC, *Developing the Technology Matrix for India and Ukraine: Draft Report,* "August 2000, p. 38.

52

METHODOLOGY ASSESSMENT: This project is incorrectly identified as an additional project by all of the approaches, except for the possibility of being correctly identified as non-additional in one set of circumstances under the U.S. approach. As discussed in a previous example, by its very nature, the U.S. percentile threshold test will qualify a certain percentage of business-as-usual projects, depending on the value that is chosen for "X." What is particularly notable in this example is that the technology matrix, which possesses a stringent additionality test in comparison to the U.S. and EU approaches, also fails to screen out this free-rider project.

Under the technology matrix approach, the test for additionality is based on an examination of the economic feasibility and market penetration of individual technologies. Technologies that are determined to be commercial based on these two tests are deemed non-additional, while non-commercial technologies are judged as additional. In this particular example, NGCC is determined to be economically feasible, but is determined to have little or no market penetration in India, and thus qualifies as additional. The market penetration test is designed to identify the existence of non-economic barriers. Non-economic barriers to implementation of a particular technology may include risks associated with installing and operating locally unknown technologies, institutional barriers or internal organizational structures that discourage investment in energy sector improvements, and poorly functioning capital markets.[12] Thus, although a project may be economic, there may exist a wealth of other reasons that it may not be implemented in a given country.

The project in question is the first NGCC project to be implemented in India. Such first-of-its-kind projects do not necessarily always come into being due to an economic subsidy such as credits. In this particular case, the project developers may have been better informed of the technology than others or have witnessed the advantages of this technology in the U.S. and simply decided to take the risk of implementing the technology in India. However, because it is applied on a country-by-country basis, the market penetration test will of necessity always qualify first-of-its-kind projects as additional. It would, of course, be possible to apply the market penetration test on a global, rather than country level. This would serve to tighten the market penetration test and consequently exclude free rider projects of this kind from gaining credits. This would come at a cost, however, as a tightened market penetration test may also exclude truly additional projects by ignoring the non-commercial barriers to implementation of a particular technology in other countries.

Thus, all of these approaches demonstrate the capacity to allow non-additional, free rider projects to attain credits. Additionally, because the approaches award credits to this free-rider project, the error in credits awarded will be 100 percent. Note, however, that the size of the error is larger for the U.S. proposal and hybrid technology matrix than for the full technology matrix. Because the U.S. and hybrid technology matrix approaches use the weighted fossil average to determine the benchmark (i.e., 1.58 lbs CO_2/kWh), while

---

[12] SAIC, "*Developing the Technology Matrix for India and Ukraine, Draft Report,*" August 2000, p. 15.

the full technology matrix approach uses the average of the most likely alternative technology--gas-fired steam turbines (i.e., 1.17 lbs CO_2/kWh), the error in credits awarded under the U.S. and hybrid technology matrix approaches will be 0.7 lbs CO_2/kWh, while the error will only be 0.29 lbs CO_2/kWh under the full technology matrix approach. In general, the technology matrix approach does a better job of matching the assumed counterfactual, and hence the benchmark, to the specific technology utilized by each project. Consequently, errors in the amount of credits awarded are lower.

PROJECT NUMBER: ES5

COUNTRY: India

SECTOR: Electricity

PROJECT TITLE: Gas Turbine Plant

PROJECT DESCRIPTION: This project involves the construction of a 10-mile distribution pipeline from the main the Uzbeki-Indian gas pipeline (first introduced in Project ES3) to the site of a planned new 10-MW gas turbine power plant. The plant will be used to meet rapidly growing peak demand in the surrounding area. Due to siting restrictions, the plant cannot be located closer to the main transmission line. Without the pipeline extension, the only fuel available at the site would be diesel fuel. The project is a joint venture between the U.S.-led consortium that owns the pipeline and the Indian utility.

PROJECT ADDITIONALITY: To meet the area's expanding peak demand, the Indian utility had to build a plant at the chosen site; the only choice it faced was whether the plant should utilize diesel generators or gas turbines. Before the potential for credits and additional financing from the pipeline company was factored in, the project economics slightly favored the diesel plant over the gas turbine plant. However, with the added incentive of the credits factored in, the economic analyses favored the gas turbine plant over the diesel generators. Hence the decision to extend the pipeline and utilize gas was determined by the incentives available through an international carbon offset program, and the project is additional.

PROJECT EMISSIONS: The power plant's heat rate is expected to be 15,000 Btus/kWh. Hence using the natural gas emissions factor of 117.1 lbs CO_2/mmBtu, the emission rate for the project can be computed as follows:

$$ER = (15{,}000 \text{ Btus/kWh})(117.1 \text{ lbs } CO_2/\text{mmBtu})(1 \text{ mmBtu}/1{,}000{,}000 \text{ Btus})$$

$$ER = 1.76 \text{ lbs } CO_2/\text{kWh}$$

PROJECT BENCHMARK DATA: We will assume that plant-level data on India's diesel generating units are unavailable. However, average data on the recently installed diesel plants, which number 200, are available. In the following table, these (fictional) aggregate data have been added to the data on coal- and gas-fired plants introduced in previous projects.

Utility	Power Plant	Prime Mover*	Fuel Type	Capacity (MW)	Year Commissioned	Heat Rate** (Btus/kWh)	Emissions Rate** (lbs CO₂/kWh)
NTPC	Talcher	ST	Coal	1000	1995	10,015	2.06
Orissa	1b Valley	ST	Coal	420	1995	10,218	2.10
Damodar Valley	Mejia	ST	Coal	630	1995-96	9,745	2.00
BSES Ltd.	Dahanu	ST	Coal	500	1995	10,610	2.18
Bihar SEB	Tenughat	ST	Coal	420	1996	10,466	2.15
Total/Average	NA	ST	Coal	2970	NA	10,211 (10,150)	2.10 (2.09)
ACME	Gas 1	ST	Gas	400	1995	10,500	1.23
ACME	Gas 2	ST	Gas	500	1997	10,100	1.18
ACME	Gas 3	ST	Gas	1000	1999	9,500	1.11
Total/Average	NA	ST	Gas	1900	NA	10,033 (9,868)	1.17 (1.15)
Small Op	Plant A	GT	Gas	10	1995	16,000	1.87
Small Op	Plant B	GT	Gas	15	1995	16,100	1.89
Small Op	Plant C	GT	Gas	5	1995	15,700	1.84
Small Op	Plant D	GT	Gas	5	1996	14,500	1.70
Small Op	Plant E	GT	Gas	10	1997	15,600	1.83
Small Op	Plant F	GT	Gas	10	1997	16,200	1.90
Small Op	Plant G	GT	Gas	10	1997	15,300	1.79
Total/Average	NA	GT	Gas	65	NA	15,629 (15,746)	1.83 (1.85)
Gas Total/Average	NA	All	Gas	1965	NA	13,950 (10,063)	1.63 (1.18)
Diesel Total/Average	NA	IC	Diesel fuel	450	1995-99	18,000 (18,000)	2.91 (2.91)
Fossil Total/Average	NA	All	All	5385	NA	17,630 (9,270)	2.83 (1.58)

*ST = Steam Turbine. GT = Gas Turbine. **Averages shown are arithmetic. Weighted averages are in parentheses.

PROJECT ANALYSIS TABLE: Project Number ES5

	U.S. Proposal	EU Proposal	Full Technology Matrix	Hybrid Technology Matrix
Does project qualify?	If 10th percentile $< X < 40^{th}$ percentile, then the project will not qualify (ER > 1.7). If $X < 10^{th}$ percentile, then more data is needed to determine additionality. However, even with the addition of more data (e.g., from surrounding countries), the project is unlikely to qualify.	The project does not qualify as additional under the EU's positive list. Gas turbines are not an advanced technology for gas-fired power plants.	The project does not involve advanced, non-commercial technology, so it will not qualify under the technology matrix. The project developers could attempt to demonstrate project additionality using the project-specific approach. However, because this is a small project, the transaction costs would likely be too high to warrant use of the project-specific approach.	The project does not qualify under the hybrid technology matrix, and the project developers would probably be unwilling to use the more costly project-specific approach for such a small project.
Is the project correctly identified as either a free rider or an additional project?	No. The project fails to qualify because the U.S. additionality determination is based on a comparison of fuel type rather than prime mover type.	No. The EU's positive list is geared more towards implementing renewable energy projects or projects of high efficiency.	No. When the project is small and thus, the transaction costs are too high to use the fallback of the project-specific approach, the technology matrix may deny some truly legitimately smaller additional projects from qualifying for credits.	No. When the project is small and thus, the transaction costs are too high to use the fallback of the project-specific approach, the technology matrix may deny some truly legitimately smaller additional projects from qualifying for credits.
Number of credits Awarded	The project does not qualify (credits = 0).	Not applicable.	The project does not qualify (credits = 0).	The project does not qualify (credits = 0).
Error in credits Awarded	Because this project is awarded no credits, but needs the incentive of credits in order to be economically feasible, diesel generators will be used instead of the gas turbine.	Not applicable.	Because this project is awarded no credits, but needs the incentive of credits in order to be economically feasible, diesel generators will be used instead of the gas turbine.	Because this project is awarded no credits, but needs the incentive of credits in order to be economically feasible, diesel generators will be used instead of the gas turbine.

METHODOLOGY ASSESSMENT: This is an additional project that fails to classify as such under each of the approaches. Under the U.S. approach, the project failed to qualify as additional when X was set between the 10th and 40th percentile. This is another example in which the U.S. approach fails because its additionality determination is based on a comparison of fuel type rather than prime mover type. In this case, despite the fact that the project involves a gas turbine plant, a percentile threshold is determined using the available data for all gas plants, including both gas turbines and steam turbines. As discussed previously, steam and gas turbines are inherently different in nature. Steam turbines are larger and more efficient than gas turbines. Because steam turbine data was included in the distribution, the average emissions rate was weighted lower than it would have been if only the gas turbine data were used to determine a threshold. In fact, if the 20th percentile using simply the gas turbine data were chosen as the threshold, then this project would have qualified as additional. Thus, this example again illustrates that the U.S. approach should be adjusted to allow for an additionality determination that considers prime mover type as well as fuel types.

Although the project fails to qualify under the full and hybrid technology matrix approaches, the project developers could attempt to demonstrate project additionality using the project-specific approach. The project-specific approach is designed to function as the fallback when the market penetration and economic feasibility tests fail to identify an additional project. However, when a project is small, the transaction costs may be too high to warrant use of the project-specific approach, thus denying some truly legitimately smaller additional projects from qualifying for credits. It is important to recognize that, although the technology matrix has the fallback of the project-specific approach, this fallback is not a panacea, and it may not be an option for developers of small projects.

In this case, the EU approach also excludes this legitimately additional project. This may not be viewed as a failure from the EU's perspective, as the positive technology list is geared more towards implementing renewable energy projects or projects of high efficiency. Gas turbine plants are lower efficiency plants; thus, from the EU perspective, the exclusion of such a project may not be deemed as a failure. However, from the U.S. perspective, this is a failure, as a legitimately additional project--one which would award credits to the U.S.--fails to be identified as such.

The implications of the failure of this project to qualify for credits are that the project would not go forward and diesel generators would be used instead of the gas turbine. This follows from the fact that the gas turbine project is truly additional, and that, as such, it requires the incentive of credits in order to be deemed feasible. The failure to qualify truly additional projects such as this one will have no net impact on the environment, but it will reduce the amount of credits awarded to the U.S. and other developed countries, thereby raising the cost of implementation.

PROJECT NUMBER: ES6

COUNTRY: India

SECTOR: Electricity

PROJECT TITLE: Wind Power

PROJECT DESCRIPTION: India is the world's fifth largest wind power generator with 1167 MW of installed capacity. However, the Indian government heavily subsidizes wind power, at significant cost. Let us suppose that, in order to reduce these subsidies, the government undertakes a program to attract foreign investment for wind power projects. The incentive offered to potential foreign investors is the credits available to renewable projects through an international carbon offset program

This project is the first undertaken as part of the new program. It involves the installation of a 10-MW wind farm in rural India. The project will produce electricity for the grid, thereby displacing fossil-generated power (whenever electricity supply and demand are in balance). The project is a joint venture between an American company and an Indian company. Both companies own a 50-percent share of the project, and all credits awarded to the project will be divided equally among the partners.

PROJECT ADDITIONALITY: Although India has an extensive wind power industry, this industry requires heavy government subsidies. The new program to attract foreign investment in effect replaces government subsidies with credits as the incentive for undertaking wind power projects. The 10-MW project, like other Indian wind power projects, would be sub-economic without either credits or government subsidies. Therefore, the project is additional.

PROJECT EMISSIONS: The project will not produce any greenhouse gas emissions.

PROJECT BENCHMARK DATA: We will assume that plant-level data on India's diesel generating units are unavailable. However, average data on the recently installed diesel plants, which number 200, are available. In the following table, these (fictional) aggregate data have been added to the data on coal- and gas-fired plants introduced in previous projects.

Utility	Power Plant	Prime Mover*	Fuel Type	Capacity (MW)	Year Commissioned	Heat Rate** (Btus/kWh)	Emissions Rate** (lbs CO_2/kWh)
NTPC	Talcher	ST	Coal	1000	1995	10,015	2.06
Orissa	1b Valley	ST	Coal	420	1995	10,218	2.10
Damodar Valley	Mejia	ST	Coal	630	1995-96	9,745	2.00
BSES Ltd.	Dahanu	ST	Coal	500	1995	10,610	2.18
Bihar SEB	Tenughat	ST	Coal	420	1996	10,466	2.15
Total/Average	**NA**	**ST**	**Coal**	**2970**	**NA**	**10,211 (10,150)**	**2.10 (2.09)**
ACME	Gas 1	ST	Gas	400	1995	10,500	1.23
ACME	Gas 2	ST	Gas	500	1997	10,100	1.18
ACME	Gas 3	ST	Gas	1000	1999	9,500	1.11
Total/Average	**NA**	**ST**	**Gas**	**1900**	**NA**	**10,033 (9,868)**	**1.17 (1.15)**
Small Op	Plant A	GT	Gas	10	1995	16,000	1.87
Small Op	Plant B	GT	Gas	15	1995	16,100	1.89
Small Op	Plant C	GT	Gas	5	1995	15,700	1.84
Small Op	Plant D	GT	Gas	5	1996	14,500	1.70
Small Op	Plant E	GT	Gas	10	1997	15,600	1.83
Small Op	Plant F	GT	Gas	10	1997	16,200	1.90
Small Op	Plant G	GT	Gas	10	1997	15,300	1.79
Total/Average	**NA**	**GT**	**Gas**	**65**	**NA**	**15,629 (15,746)**	**1.83 (1.85)**
Gas Tot/Average	NA	All	Gas	1965	NA	13,950 (10,063)	1.63 (1.18)
Diesel Total/Average	NA	IC	Diesel fuel	450	1995-99	18,000 (18,000)	2.91 (2.91)
Fossil Total/Average	NA	All	All	5385	NA	17,630 (9,270)	2.83 (1.58)
Sector Total/Average	NA	All	All	5385	NA	17,630 (9,270)	2.83 (1.58)

*ST = Steam Turbine. GT = Gas Turbine. **Averages shown are arithmetic. Weighted averages are in parentheses.

60

PROJECT ANALYSIS TABLE: Project Number ES6

	U.S. Proposal	EU Proposal	Full Technology Matrix	Hybrid Technology Matrix
Does project qualify?	Because this is a zero emissions project, it automatically qualifies as additional. Under the U.S. proposal, for a zero emissions project, there is no percentile test or eligibility threshold.	This project qualifies as additional under the EU's positive list. It falls under the category of renewable energy -- i.e., wind.	This project qualifies as additional under the technology matrix, because even in the face of capital cost competitiveness and significant market penetration, wind power technology requires some kind of economic incentive or favorable financing. It is therefore considered an advanced, non-commercial technology.	Project will automatically qualify as additional.
Is the project correctly identified as either a free rider or an additional project?	Yes.	Yes.	Yes.	Yes.
Number of credits Awarded	Under the U.S. proposal, a zero emissions project must use the weighted sector average for recently built capacity to calculate credits. Thus, the number of credits awarded would be 1.53-0=1.53 lbs CO_2/kWh.	Not applicable.	For wind turbine technology in India, it has previously been determined that the benchmark should represent a sector average for all recently built capacity.* Thus, the credits awarded would be 2.83-0=2.83 lbs CO_2/kWh.	The number of credits awarded would be determined by using the weighted sector average for recently built capacity. Thus, the credits awarded would be 1.53-0=1.53 lbs CO2/kWh.
Error in credits Awarded	Unknown.	Not applicable.	Unknown.	Unknown.

*SAIC, *Developing the Technology Matrix for India and Ukraine, Draft Report,* August 2000, Table 5, p. 38.

METHODOLOGY ASSESSMENT: This project is correctly identified as additional under all four approaches. This particular example demonstrates that all four approaches are well adapted to identifying additional renewable energy projects. This is the first example of an additional project that was successfully identified as such under all approaches.

In this particular case, it is apparent that the U.S. and hybrid technology matrix approaches award fewer credits than the technology matrix approach because these approaches use a benchmark based on the weighted versus the arithmetic average emissions rate. Whether the use of a weighted average versus an arithmetic average yields a more accurate determination of credits is dependent on the particular circumstances surrounding the project in question. For example, if a baseload facility is compared against data consisting mostly of baseload facilities, then the points of comparison are similar enough for an arithmetic average to yield an accurate determination of credits. Usually, under the technology matrix approach, like facilities are compared to each other. However, a problem arises when, under the technology matrix approach, a project is compared to a data set consisting of a mixture of different facility types. Peaking facilities are much more abundant, smaller, and less efficient than baseload facilities. If a baseload facility is compared to data consisting of a mixture of peaking and baseload facilities, then the peaking facilities, due to their volume, will tend to dominate the arithmetic average for the facility data set. In this particular case study, the technology matrix approach requires the use of a sector average, and since this is an arithmetic average, the small peaking units distort the resulting baseline. To enable it to better handle projects such as this one, the technology matrix approach should be revised to utilize a weighted, rather than arithmetic, average.

PROJECT NUMBER: ES7

COUNTRY: Kazakhstan

SECTOR: Electricity

PROJECT TITLE: IGCC in Kazakhstan

PROJECT DESCRIPTION: A decade after the breakup of the Soviet Union, GDP and electricity demand remains well below 1990 levels throughout the FSU. Hence, unlike in India, there is little need for new power plants to meet new demand; existing capacity is more than sufficient to meet demand. However, there is a critical need to modernize and rebuild existing power plants. In many cases these plants are old and obsolete, and have deteriorated dramatically during the economic crisis of the 1990s. Kazakhstan, like other countries of the FSU, is working to refurbish its aging power plants.

However, in some cases the existing power plants are simply to obsolete and worn out to warrant further investment. Let us suppose that Kazakhstan is seeking the foreign capital required to replace, rather than rebuild, some of its existing power plants. This project is the first in what the government hopes will be a series of U.S.-backed joint ventures in the Kazak power generation sector. The project involves the construction of a new, 400-MW IGCC power plant. The plant will utilize coal as its primary fuel. The U.S. company is providing significant financing for the project, in exchange for an ownership share and all of the credits to be awarded to the project. No existing plants will be shut down because of the project, due to the need to maintain employment levels. However, generation from the new IGCC power plant will enable Kazakhstan to reduce generation from its older, much less efficient existing facilities.

PROJECT ADDITIONALITY: The American company was prepared to invest in the Kazak power sector with or without an international carbon offset program. However, the added incentive of credits was the key factor in the decision to utilize IGCC technology rather than conventional coal-fired technology. At present, IGCC is an advanced, non-commercial technology, and it requires subsidies or other aid in order to be competitive with conventional technology. Therefore, the project is additional.

PROJECT EMISSIONS: The power plant is expected to operate with an average heat rate of 7560 Btus/kWh. Based on an emission factor of 205.3 lbs CO_2/mmBtu for bituminous coal, the project's emission rate (ER) is estimated as follows:

$$ER = (7560 \text{ Btus/kWh})(205.3 \text{ lbs } CO_2/\text{mmBtu})(1 \text{ mmBtu}/1{,}000{,}000 \text{ Btus})$$

$$ER = 1.55 \text{ lbs } CO_2/\text{kWh}$$

PROJECT BENCHMARKS: As a result of the economic collapse following the breakup of the Soviet Union, no new power plants have been built in Kazakhstan or in the surrounding FSU countries in recent years. Therefore, no data are available to support the development of either country-specific or regional benchmarks.

63

PROJECT ANALYSIS TABLE: Project Number ES7

	U.S. Proposal	EU Proposal	Full Technology Matrix	Hybrid Technology Matrix
Does the project qualify?	The data required to establish a threshold is unavailable either for Kazakhstan or for other FSU countries. In this case, it might be necessary to use either a continental or even global threshold to establish additionality. However, since IGCC is significantly more efficient than conventional coal-fired technology, it is likely to qualify regardless of the data used.	This project does not qualify under EU's positive list, because the technology would be coal-fired, and thus does not fall into the categories of renewables, energy efficiency, or demand side management.	IGCC represents an advanced, non-commercial technology, and thus the project qualifies as additional.	This project qualifies as additional under the Hybrid Technology Matrix.
Is the project correctly identified as either a free rider or an additional project?	Yes.	No. The project involves a coal-fired plant, and the EU's positive list aims to steer countries away from coal and other fossil fuels, and towards renewables or more efficient technologies.	Yes.	Yes.
Number of credits Awarded	Difficult to determine. The benchmark would have to be based on global or larger-region coal-fired power plant data.	Not applicable.	Difficult to determine. The benchmark would have to be based on global or larger-region coal-fired power plant data.	Difficult to determine. The benchmark would have to be based on global or larger-region coal-fired power plant data.
Error in credits Awarded	Unknown.	Not applicable.	Unknown.	Unknown.

METHODOLOGY ASSESSMENT: The only approach under which this project was not correctly identified as additional was the EU approach. Because the project is coal-fired, it does not fall under any of the technology categories in the EU positive list. The goal under the EU approach is to steer developing countries away from coal and other fossil fuels, and towards renewables, or at the very least, more efficient technologies than coal-fired technologies. Interestingly, while the EU has not issued a formal test for additionality, many NGOs, both in Europe and in the US, have maintained that the EU's positive list is an adequate and favorable test for additionality. However, in this example, the EU's positive list fails to identify a legitimately additional project. For the power sector, the EU's positive list offers a very stringent test for additionality, but it fails to qualify technologies such as IGCC, which is a non-conventional, more efficient technology that is likely to be additional in many countries. From the U.S. perspective, the EU's positive list is not a successful test of additionality, especially considering that the U.S. has clearly indicated a desire to include certain coal-fired technologies under the CDM or similar flexible, market-based flexible mechanisms or program. Such technologies will clearly fail to qualify as additional under the EU's positive list.

This project represents the problem that arises when certain countries or areas lack the comparable facilities required to support the development of country-specific or regional benchmarks. In this particular case, due to the economic collapse following the breakup of the Soviet Union, no new power plants have been built in the country of reference (i.e., Kazakhstan) or surrounding FSU countries in recent years. Data requirements to determine additionality are only an issue under the U.S. approach, which requires enough data to define a percentile distribution. However, the determination of a benchmark is problematic under the technology matrix approaches as well. Presumably, the benchmark would have to be based on global or regional coal-fired power plant data.

PROJECT NUMBER: ES8

COUNTRY: Tajikistan

SECTOR: Electricity

PROJECT TITLE: Hydropower

PROJECT DESCRIPTION: Although it is the poorest of the former Soviet republics, Tajikistan nonetheless possesses a significant aluminum industry based on the country's abundant hydropower resources. This project involves the construction of a dam and 2000-MW hydropower plant. The plant will provide low-cost electricity for a new aluminum production facility; in addition, a portion of the power generated by the plant will be exported to neighboring countries.

An American company will provide the bulk of the financing for the aluminum plant, dam, and hydropower station, in partnership with a local firm. Any credits generated by the project will be retained by the U.S. company.

PROJECT ADDITIONALITY: This is a large-scale project, made risky by the political and economic instability characterizing Tajikistan. Relative to the magnitude of the risks and capital investment required, the impact of the credits on the project's "bottom line" is insignificant. In other words, an international carbon offset program would not provide incentives sufficient to warrant project implementation; rather the project stands on its own economic merits. It is therefore not additional.

PROJECT EMISSIONS: The hydropower plant will not produce any greenhouse gas emissions.

PROJECT BENCHMARKS: As a small, poor nation, Tajikistan is unlikely to attract a large number of carbon offset program-related projects. Furthermore, it lacks the resources necessary to develop country-specific benchmarks. Data are also lacking for the countries surrounding Tajikistan. In short, no data are available to support the development of either country-specific or regional benchmarks.

PROJECT ANALYSIS TABLE: Project Number ES8

	U.S. Proposal	EU Proposal	Full Technology Matrix	Hybrid Technology Matrix
Does project qualify?	As a zero emissions project, it is not subject to the percentile test or eligibility threshold and therefore the project automatically qualifies.	The positive list only allows for small-scale hydropower projects. As a large-scale hydro project, it would not qualify.	The project does not involve advanced, non-commercial technology and will not qualify under the technology matrix. Project developers may use the project-specific approach to determine additionality; however, because the project is economical, it is unlikely to qualify under project-specific.	The project does not involve advanced, non-commercial technology and will not qualify under the technology matrix.
Is the project correctly identified as either a free rider or an additional project?	No. The percentile test can eliminate most free riding conventional technology fossil fuel projects, but it is defenseless against free riding, commercial zero emissions projects.	Yes	Yes	Yes
Number of credits Awarded	A power sector average would be required to calculate credits for this project, but this information is unavailable on a national and regional basis. Since Tajikistan is an FSU country, it may be reasonable to use a sector average from another FSU country (e.g. Ukraine or Russia) where data would be available. However, there may not be any "recent" plants, as required by the U.S. proposal, built in either country from which to construct the data. Also, translating Russian or Ukrainian data for use in hydro-dominated Tajikistan may be erroneous.	Not applicable	The project does not qualify for credits	The project does not qualify for credits
Error in credits Awarded	Unknown	Not applicable	The error in credits is zero	The error in credits is zero

METHODOLOGY ASSESSMENT: This non-additional project is correctly identified as such under the EU positive list proposal and the technology matrix approach, but under the U.S. proposal, it is incorrectly identified as additional and would qualify for credits. When considering large-scale hydro projects, the EU proposal and the technology matrix will always identify them as non-additional because the EU's positive list only includes small-scale hydro while the technology matrix only allows for advanced, non-commercial technologies. In contrast, hydro projects (small or large-scale) under the U.S. proposal will always qualify as additional because as a zero emissions project, hydro is not subject to the percentile test and automatically qualifies as additional. Unlike fossil fuel projects where the percentile test can eliminate most non-additional, conventional technology projects, the U.S. proposal is defenseless against non-additional, commercial zero emissions projects which include not only hydro, but nuclear and renewables as well. The U.S. proposal will not eliminate any non-additional, zero emissions project from qualifying for credits. For renewables, this automatic pass to credits is fine in the sense that virtually all non-hydro renewable projects will utilize advanced non-commercial technologies that are truly additional and credits are likely to be an important factor in project developers moving forward with a renewable project. However, hydro and nuclear technologies have been developed commercially worldwide. Moreover, the capital costs for these technologies are very large and potential credits are highly unlikely to tip the economic scales of a hydro or nuclear project. In addition, hydro projects often are built for reasons other than electricity generation (e.g. flood control, reservoir development, irrigation, etc.). Many times the electricity generation is ancillary to the project's main purpose which further strengthens the point that most hydropower projects are likely to be business as usual or non-additional.

This hydropower project also illustrates a much larger issue. As noted, no data are available on either a national or a regional basis to support the development of benchmarks, meaning that at this point it is impossible to develop the sector average that would be used to calculate credits. One option to deal with this problem would be to use a sector average from another FSU country, such as Russia or Ukraine where data would be available. Unfortunately, there are two problems with this approach: 1) the U.S. proposal requires a comparison to "recent" facilities, and there may not be any recently constructed plants in any FSU countries from which to construct the data; and 2) if recent plant data from other FSU countries were available, translating that data to hydro-dominated Tajikistan may be fraught with error. This lack of data is a problem that project developers are likely to encounter throughout the developing world. Many developing countries currently lack the necessary data to support emission reduction projects and lack funding to support data collection efforts. Unless this data problem is addressed prior to the initiation of an international carbon offset program, project developers will be faced with the expense of collecting the data themselves, which in turn would significantly add to project costs. Moreover, if project developers are left to collect the data, the objectivity and accuracy of the data could be called into question. It is important to emphasize that, although this issue arises in the context of the U.S. proposal in this particular example, data deficiencies will in fact affect baseline development under all four methodologies. This follows from the fact that, once a project qualifies for credits, all four methodologies require sector-wide data upon which to estimate the benchmark.

PROJECT NUMBER: ES9

COUNTRY: India

SECTOR: Electricity

PROJECT TITLE: Distributed Generation: Fuel Cells

PROJECT DESCRIPTION: As is the case in many developing countries, India's electricity transmission and distribution system is plagued with high line losses and outright theft. Furthermore, there is a pressing need for new transmission capacity, both to provide service in remotely located areas and to keep pace with demand in high growth areas. Distributed generation offers India an alternative to transmission capacity expansion (which requires large capital investments) as well as a means of reducing line losses.

Diesel generators can be used in a distributed mode, but there are disadvantages to this approach, including the relatively low efficiency of diesel generators and the difficulties and costs of delivering diesel fuel to remote areas in India. Fuel cells, which are highly efficient and utilize natural gas, represent a potential solution to these problems. Let us suppose that the fictional Uzbeki-Indian natural gas pipeline, introduced in previous projects, provides a source of natural gas that could be used to supply fuel cells in some areas of western India. This project involves the installation of a 2-MW fuel cell unit, to provide electricity in one such area. The fuel cell unit will be located near the center of a group of villages that have been experiencing rapid growth. The transmission network supplying the villages is already operating near capacity. The fuel cell installation is a pilot project which, if successful, will be duplicated in other locations in and around the villages, as an alternative to expanding the transmission network. The project is a joint venture between an U.S. investor-owned utility and an Indian IPP. Any credits awarded to the project will be split amongst the partners.

PROJECT ADDITIONALITY: Fuel cell technology remains non-commercial and experimental at this time, and hence the project involves considerable risk. Without the credits, the project is considered sub-economic. However, the prospect of obtaining credits offsets the relatively high capital costs associated with fuel cells, and renders both the pilot project, and the possible expansion of the project to commercial scale, economically feasible. Thus, the project is not a free rider.

PROJECT EMISSIONS: The project will not produce any greenhouse gas emissions.

PROJECT BENCHMARK DATA: The following data on India's recently installed power plants, introduced in previous projects, are available for benchmark development.

Utility	Power Plant	Prime Mover*	Fuel Type	Capacity (MW)	Year Commissioned	Heat Rate** (Btus/kWh)	Emissions Rate** (lbs CO$_2$/kWh)
NTPC	Talcher	ST	Coal	1000	1995	10,015	2.06
Orissa	1b Valley	ST	Coal	420	1995	10,218	2.10
Damodar Valley	Mejia	ST	Coal	630	1995-96	9,745	2.00
BSES Ltd.	Dahanu	ST	Coal	500	1995	10,610	2.18
Bihar SEB	Tenughat	ST	Coal	420	1996	10,466	2.15
Total/Average	NA	ST	Coal	2970	NA	10,211 (10,150)	2.10 (2.09)
ACME	Gas 1	ST	Gas	400	1995	10,500	1.23
ACME	Gas 2	ST	Gas	500	1997	10,100	1.18
ACME	Gas 3	ST	Gas	1000	1999	9,500	1.11
Total/Average	NA	ST	Gas	1900	NA	10,033 (9,868)	1.17 (1.15)
Small Op	Plant A	GT	Gas	10	1995	16,000	1.87
Small Op	Plant B	GT	Gas	15	1995	16,100	1.89
Small Op	Plant C	GT	Gas	5	1995	15,700	1.84
Small Op	Plant D	GT	Gas	5	1996	14,500	1.70
Small Op	Plant E	GT	Gas	10	1997	15,600	1.83
Small Op	Plant F	GT	Gas	10	1997	16,200	1.90
Small Op	Plant G	GT	Gas	10	1997	15,300	1.79
Total/Average	NA	GT	Gas	65	NA	15,629 (15,746)	1.83 (1.85)
Gas Total/Average	NA	All	Gas	1965	NA	13,950 (10,063)	1.63 (1.18)
Diesel Total/Average	NA	IC	Diesel fuel	450	1995-99	18,000 (18,000)	2.91 (2.91)
Fossil Total/Average	NA	All	All	5385	NA	17,630 (9,270)	2.83 (1.58)
Sector Total/Average	NA	All	All	5385	NA	17,630 (9,270)	2.83 (1.58)

*ST = Steam Turbine; GT = Gas Turbine; IC = Internal Combustion, **Averages shown are arithmetic. Weighted averages are in parentheses.

PROJECT ANALYSIS TABLE: Project Number ES9

	U.S. Proposal	EU Proposal	Full Technology Matrix	Hybrid Technology Matrix
Does project qualify?	As a zero emissions project, it is not subject to the percentile test or eligibility threshold and therefore the project automatically qualifies.	Although the positive list allows for advanced technologies, it does not include a category that would cover distributed generation fuel cells; therefore, the project would not qualify.	Fuel cells are an advanced non-commercial technology. Projects using this technology will automatically qualify.	Fuel cells are an advanced non-commercial technology. Projects using this technology will automatically qualify.
Is the project correctly identified as either a free rider or an additional project?	Yes	No. There is no mention of fuel cells as a qualifying technology anywhere within the positive list.	Yes	Yes
Number of credits Awarded	Under the U.S. proposal, a zero emissions project must use a sector average to calculate credits. For this example, the sector average is 2.83 lbs CO_2/kWh. The estimated credits would be 2.83 – 0=lbs CO_2/kWh.	Not applicable	In a previous report, * it was determined that for distributed generation fuel cell projects, project developers should use the project specific approach for baseline development. Because the most likely alternative to the project will depend on the economic specifics of the project, baseline development using the project specific approach, rather than the technology matrix approach is likely to yield a more accurate emission reduction estimate. In this case, the likely alternative is expanding the transmission network so a fossil average would be used. The estimated credits would be 2.83 – 0 = 2.83 lbs CO_2/kWh.	Estimated credits: 2.83 – 0=2.83 lbs CO_2/kWh
Error in credits Awarded	Unknown	Not applicable	Unknown	Unknown

*SAIC, "*Developing the Technology Matrix for India and Ukraine: Draft Report*," August 2000, pp.64-65.

METHODOLOGY ASSESSMENT: This additional project will qualify for credits under all of the proposals except for the EU proposal. There is no mention of fuel cells as a qualifying technology anywhere in the positive list. Whether this omission is intentional or merely an oversight is not clear at the moment; the EU proposal has not been refined since it was first introduced in September 2000. This circumstance most certainly needs clarification prior to an operational international carbon offset program, similar to the CDM. Moreover, if the project were disqualified under the positive list, the likely alternative would be expansion of the current transmission network or perhaps diesel generator in a distributed mode, a scenario the positive list attempts to discourage.

Under the technology matrix approach, credits will be awarded at a rate of 2.83 lbs CO_2/kWh generated using the project specific approach to develop the baseline. In a previous report, it was determined that distributed generation fuel cell projects should use the technology matrix approach to determine additionality, but then the project specific approach is used for baseline development.[13] In this type of project, because the most likely alternative to the project will depend on the economic specifics of the project, a combined approach will yield a more accurate emission reduction estimate.

Applying the project-specific approach, the fossil average was used to calculate credits. However, use of the fossil average is a significant simplification of the emission reduction estimate. It is likely that the increased generation would come from marginal units; therefore, an even more accurate emission reduction estimate would result from the average emissions rate of marginal units, weighted by the amount of time each of these units operate at the margin. Calculating a marginal average would require production cost models and much more time and effort would be required to produce the average. At this point, baseline development would likely become cost prohibitive in light of the fact that this is a small-scale project (2MW). This, in turn, would decrease the likelihood of the project developers moving forward with the project. The costs associated with project-specific baseline development will always be an issue, especially in the case of small-scale projects. However, it is important to emphasize that this project is treated in an unusual manner under the technology matrix approach. Normally, a standard benchmark will be made available for the computing of credits. Fuel cell technology for distributed generation applications is one of the few cases for which a project-specific approach has been recommended under the technology matrix.

---

[13] SAIC, *"Developing the Technology Matrix for India and Ukraine: Draft Report,"* August 2000, pp.64-65.

PROJECT NUMBER: ES10

COUNTRY: China

SECTOR: Electricity

PROJECT TITLE: Transmission Capacity Expansion

PROJECT DESCRIPTION: China, like India, is experiencing rapid growth in its population and economy. This growth, in turn has placed pressure on the country's transmission and distribution network, which in many cases is operating at or near capacity. This project involves the replacement of an existing transmission line, currently operating at capacity, with a larger-diameter, higher voltage line. The project is being undertaken primarily to increase the capacity of the line, but the project will also have the secondary effect of reducing line losses. The project is being undertaken by the local Chinese utility. However, some limited financing is being provided by an American company, in exchange for the credits awarded to the project.

PROJECT ADDITIONALITY: This project is a free rider. The Chinese utility must and will undertake the project in order to increase the line's capacity. However, seeing an opportunity to obtain additional project financing, the utility is selling the credits to the American company.

PROJECT EMISSIONS: The project is expected to reduce the transmission line's losses from 0.1 percent per mile to 0.05 percent per mile. It has been determined that the average emissions rate for the generation of electricity carried by the line is 1.8 lbs CO_2/kWh. Therefore, the project's emissions rate (ER) can be estimated as follows:

$$ER = (0.0005 \text{ kWh lost/kWh transmitted/mile})(1.8 \text{ lbs } CO_2/\text{kWh})$$

$$ER = 0.09 \text{ lbs } CO_2/\text{kWh-mile}$$

PROJECT BENCHMARKS: Let us assume that ten transmission lines, using the same standard wire size as the project, have been installed in China in the past 5 years. Estimated losses for these ten lines are as follows:

Transmission Line	Emissions Rate (lbs CO2/kWh)	Losses (Percent per Mile)	Emissions Rate (lbs CO2/kWh-mile)
Beijing 1	1.8	0.053	0.000954
Beijing 2	1.8	0.067	0.001206
Beijing 3	1.8	0.061	0.001098
Canton 1	2.0	0.059	0.001180
Canton 2	2.0	0.055	0.001100
Shanghai 1	2.1	0.051	0.001071
Shanghai 2	2.1	0.079	0.001659
Fushun	1.9	0.073	0.001387
Wuhan	2.2	0.068	0.001496
Sian	1.9	0.083	0.001577
Average	**1.96**	**0.065**	**0.0012728**

PROJECT ANALYSIS TABLE: Project Number ES10

	U.S. Proposal	EU Proposal	Full Technology Matrix	Hybrid Technology Matrix
Does project qualify?	If "X" > 10th percentile the threshold would be at least 0.000954 lbs. CO_2/kWh-mile. The project emissions rate of 0.0009 lbs CO_2/kWh-mile falls below the threshold. Therefore, the project would qualify. If "X" < 10th percentile, the available data are insufficient to establish the threshold.	The positive list allows for energy efficiency projects that significantly improve energy transmission, but the positive list does not define the term "significantly;" therefore, it cannot be determined whether or not the project would qualify.	The project is not using an advanced non-commercial technology; therefore, it would not qualify under the technology matrix. The project developers would be afforded the opportunity to demonstrate additionality under the project-specific approach but since it is a business as usual project it is unlikely to qualify.	The project will not qualify under the technology matrix and is unlikely to qualify as additional under the project-specific approach.
Is the project correctly identified as either a free rider or an additional project?	No. By only requiring projects to be "significantly better than average," the test guarantees that a certain number of conventional, business as usual projects, using commercial technologies will qualify for credits.	Indeterminate. Absent of a definition of the term "significantly," it cannot be determined whether this project would qualify for credits.	Yes	Yes
Number of credits Awarded	The benchmark would be taken as the average of the ten transmission lines (0.0012728). The estimated credits would then be 0.0012728-0.0009 = 0.0003728 lbsCO_2/kWh-mile	Not applicable	Project does not qualify for any credits	Project does not qualify for any credits
Error in credits Awarded	The project is a free rider; therefore, the error is equal to 100 percent of the credits awarded (0.0003728 lbs CO_2/kWh-mile)	Not applicable	The project is correctly identified as a free rider; therefore, the error in the credits awarded is zero.	The project is correctly identified as a free rider; therefore, the error in the credits awarded is zero.

METHODOLOGY ASSESSMENT: This free rider project is correctly identified as such only under the technology matrix approaches. The U.S. proposal incorrectly identifies it as an additional project, and it is indeterminate under the EU proposal.

This project clearly demonstrates a fundamental problem with the U.S. proposal's percentile test, i.e. that a certain number of non-additional/free rider projects will qualify for credits. Even if the percentile test is set at a stringent level (e.g. the 10th percentile), it will not effectively screen out all free rider projects. By only requiring emission reduction projects to be "significantly better than the average," the U.S. percentile test guarantees that a certain number of conventional, business as usual projects, using commercial technologies such as coal, oil, or natural gas, will qualify for credits. The percentile test will not necessarily encourage use of the most advanced and most efficient technologies, but may merely encourage use of the best of what is already commercially available. In other words, projects that pass the percentile test are not the best of the best but are merely the best of what is already out there, assuming, of course, that "X" is set at a stringent level. Because the project is not identified as a free rider, the error in credits awarded is equal to 100 percent (0.0003728 lbs CO_2/kWh-mile). It is also important to note that if "X" < the 10th percentile, the available data would be insufficient to establish a threshold.

The EU's positive list allows for energy efficiency projects that significantly improve energy transmission, but in the absence of a definition of the term "significantly," it cannot be determined whether or not the project would qualify as additional. This circumstance points to the fact that the EU needs to clarify and further refine its proposal if it is to be seriously considered as a legitimate qualification screen for an international carbon offset program.

PROJECT NUMBER: ES11

COUNTRY: India

SECTOR: Electricity

PROJECT TITLE: Carbon Sequestration Technology for an IGCC Power Plant

PROJECT DESCRIPTION: This project involves the construction of a new, 500-MW coal-fired IGCC power plant in India that is equipped with an advanced energy transfer system with which to sequester carbon. The system involves the replacement of a standard IGCC plant's combustor with fluidized bed oxidation and reduction reactors. These reactors encompass a new technology in which the gasified coal transfers its energy to reduce a metal oxide, producing high pressure CO_2 and water, and transferring its chemical energy to the metal. The steam and CO_2 are cooled and the steam is condensed. The steam drives a cycle and produces a pure stream of high pressure CO_2 that can be sequestered with little additional compression energy and stored in a natural geologic formation. The metal is then re-oxidized in air, producing heat and heating a high-pressure stream of air to drive the cycle. The metal transfers the fuel energy to the air without carrying the fuel's CO_2 along with it.

The power plant is needed to meet India's rapidly growing demand for electricity, and will operate as a baseload facility. The plant will be built as a joint venture between an Indian utility and an U.S. investor-owned utility. The U.S. utility will receive all of the credits awarded to the project, along with a share of the project's ownership, in exchange for its financial backing.

PROJECT ADDITIONALITY: Currently, coal-fired IGCC for power generation is an advanced combustion technology that is not being used on a commercial basis, either in India or elsewhere (Note, however, that oil-fired IGCC is being utilized, particularly for applications at petroleum refineries). Likewise, the proposed energy transfer carbon sequestration system technology to be added to the plant has not been used on a commercial basis anywhere in the world. The project developers initially decided to use IGCC rather than conventional technology in order to obtain the credits that would be available to an advanced-technology project under an international carbon offset program. The project developers then decided to equip the IGCC plant with the advanced energy transfer carbon sequestration system in order to obtain more credits than what would be awarded to the plant without the advanced carbon sequestration system technology. Therefore, the project is additional.

PROJECT EMISSIONS: Without being equipped with the advanced energy transfer carbon sequestration system, the IGCC power plant is expected to operate with an average heat rate of 7560 Btus/kWh. Based on an emission factor of 205.3 lbs CO_2/mmBtu for bituminous coal, the project's emission rate (ER) is estimated as follows:

77

$$ER = (7560 \text{ Btus/kWh})(205.3 \text{ lbs CO}_2/\text{mmBtu})(1 \text{ mmBtu}/1{,}000{,}000 \text{ Btus})$$

$$ER = 1.55 \text{ lbs CO}_2/\text{kWh}$$

With the addition of the proposed energy transfer carbon sequestration system, the IGCC plant's CO_2 emissions will be reduced dramatically. The improved process is expected to reduce the CO_2 emissions of the IGCC plant by 83% while suffering only a 1.5 to 4% efficiency penalty. Thus, the ER of the plant with the addition of the energy transfer carbon sequestration system is estimated as follows:

$$ER = (1.55 \text{ lbs CO}_2/\text{kWh})(.83) = 1.29 \text{ lbs CO}_2/\text{kWh}$$

$$ER = 1.55 \text{ lbs CO}_2/\text{kWh} - 1.29 \text{ lbs CO}_2/\text{kWh}$$

$$ER = 0.26 \text{ lbs CO}_2/\text{kWh}$$

PROJECT BENCHMARK DATA: A total of seven coal-fired power plants have been opened in India since 1995. The EPA has collected heat rate data on these and older power plants. Two of the post 1995 power plants have heat rate data that are suspect: 7365 and 5611 Btus/kWh. These two power plants were therefore eliminated as outliers. The heat rate data for the remaining 5 power plants are as follows:

Utility	Power Plant	Capacity (MW)	Year Commissioned	Heat Rate* (Btus/kWh)	Emissions Rate* (lbs CO_2/kWh)
NTPC	Talcher	1000	1995	10,015	2.06
Orissa	1b Valley	420	1995	10,218	2.10
Damodar Valley	Mejia	630	1995-96	9,745	2.00
BSES Ltd.	Dahanu	500	1995	10,610	2.18
Bihar SEB	Tenughat	420	1996	10,466	2.15
Total/Average	NA	2970	NA	10,211 (10,150)	2.10 (2.09)

*Averages shown are arithmetic. Weighted averages are in parentheses.

PROJECT ANALYSIS TABLE: Project Number ES11

	U.S. Proposal	EU Proposal	Full Technology Matrix	Hybrid Technology Matrix
Does project qualify?	If "X" > 20th percentile, then the threshold will be at least 2.00, and the project will qualify (ER<2.00). If X < 20th percentile then threshold test must be based on regional rather than Indian data, but it is likely that project will qualify regardless of data used (because project ER is much less than conventional coal ER).	Clean coal projects will not qualify under the positive list unless they have efficiencies > 55 percent. This project's efficiency is less than 45 percent, so it will not qualify for credits	Not only is IGCC, in and of itself, an advanced, non-commercial technology, but the proposed energy transfer carbon sequestration system is also an advanced, non-commercial technology. Projects using this technology will automatically qualify as additional under the technology matrix.	Project will automatically qualify as additional.
Is the project correctly identified as either a free rider or an additional project?	Yes	No. Project is a coal project, and such projects are not included under the EU's positive list.	Yes	Yes
Number of credits Awarded	The benchmark would be taken as either the weighted average of the five Indian plants (2.09) or the average of a larger set of regional plants. In the former case, the estimated credits would be 2.09-0.26 = 1.83 lbs/kWh	Not applicable.	Estimated credits = $2.10 - 0.26 = 1.84$ lbs/kWh	$2.10 - 0.26 = 1.84$ lbs/kWh
Error in credits Awarded	Unknown.	Not applicable.	Unknown.	Unknown.

METHODOLOGY ASSESSMENT: This additional project will qualify for credits under all approaches except the EU approach. This project is unique in that it equips an advanced technology (IGCC) with another advanced technology (the energy transfer carbon sequestration system) for the purpose of receiving more credit than that which would be received by employing just one advanced technology (IGCC) on its own. In this case, the addition of the carbon sequestration system will reduce the CO_2 emissions of the IGCC plant by 83%. While the IGCC plant, without the added carbon sequestration technology, would prove to be additional under the U.S. and technology matrix approaches (see ES1), a substantial difference in the number of credits awarded to this project will occur when the carbon sequestration system is included.

Under the full technology matrix approach, the project will be awarded credits at the rate of 1.84 lbs per kWh generated. This reduction estimate is based on a benchmark reflecting the average emission rate of all recently built coal-fired power plants in India. If IGCC were employed without the addition of the carbon sequestration system, credits would be awarded at a rate of 0.55 lbs/kWh under the technology matrix approaches (see ES1). Thus, the number of credits awarded to this additional project, with the addition of the carbon sequestration technology, is over three times the amount of credits that would have been awarded to the IGCC project without this added advanced technology.

The official U.S. approach uses a benchmark based on the weighted average emission rate, and awards credits at a rate of 1.83 lbs per kWh generated, but *only* if "X" in the percentile threshold test were set equal to or greater than the 20th percentile. If X < 20th percentile, it is necessary to define both the threshold and the benchmark on the basis of regional data (perhaps, e.g., including data for China) rather than data specific to India. The use of a sector average from another country is problematic, however. First, the U.S. proposal requires a comparison to "recent" facilities, and there may not be any recently constructed plants in any surrounding countries from which to construct the data. Secondly, if recent data from other surrounding countries were available, translating that data to another country may prove erroneous. As shown in previous case studies, this example illustrates the fact that data requirements are more difficult to meet under the official U.S. approach than under the technology matrix approach because the U.S. approach requires sufficient data to define a percentile distribution. Under the U.S. approach, if the project were employed without the addition of the carbon sequestration system, credits would be awarded at a rate of 0.54 lbs/kWh. Similar to the technology matrix approach, the amount of credits awarded to the project with the inclusion of the advanced carbon sequestration technology is over three times the amount of credits awarded to the IGCC project without the addition of this advanced technology.

As discussed in ES1, under the EU positive list of technologies, the project will fail to qualify not because it is deemed non-additional, but because it employs the use of coal. The implied goal under the EU approach is to promote the use of renewables over coal and other fossil fuels. However, as previously mentioned, it is unlikely that the project developers would opt for renewables in lieu of IGCC. The power plant is being built to meet a specific identified market need—i.e., a need for a large capacity (500-MW) baseload plant to serve rapidly growing demand. Renewable technologies such as solar and wind cannot be used for such large-capacity, baseload applications. If disqualified under the EU positive list test, it is likely that the project developers would build a conventional coal-fired plant in place of the IGCC plant. Rarely would disqualification under the positive list cause project developers to opt for a renewables plant in lieu of a fossil fuel plant, because renewables technology, even if feasible at or near the project site, serve a different market application. In short, the EU's goal of changing clean development paths via the positive list is not likely to prove successful.

PROJECT NUMBER: IS1

COUNTRY: Azerbaijan

SECTOR: Industrial

PROJECT TITLE: Installation of District Heating System

PROJECT DESCRIPTION: In many cities in Eastern Europe and the FSU, district heating systems are used to provide steam heat to apartments, hotels, and other commercial buildings. Let us assume that, due to economic expansion in the oil sector, a mid-sized city in Azerbaijan has recently been experiencing rapid growth. Specifically, a number of new hotels and high-rise apartments are being built in a concentrated area in the outskirts of the city, to accommodate the influx of foreign businessmen and oil industry workers.

The builders are considering three options for heating the new buildings: natural gas, electricity, or steam heat to be provided by a central district heating plant. Due to gas supply problems, Azerbaijan has become increasingly dependent on foreign (mostly Turkmen) imports. As a result of price increases and payment problems, imports were cutback on at least one occasion in the recent past, leaving most of the country without gas. Furthermore, for at least the next decade associated gas production from Azerbaijan's offshore oil fields is expected to be insufficient to offset the continuing production decline at the existing gas-condensate fields. Thus, faced with the prospect of continued reliance on uncertain foreign gas supplies, the builders have ruled out natural gas as an option.

Electricity is also considered an uncertain option. Due to power plant and grid maintenance problems, the city has been experiencing a significant rise in blackouts and brownouts. Since security of heat supply is considered a top priority for hotels and apartments serving foreign businessmen, the builders have decided against both gas and electricity, in favor of an oil-fired district heating system. Furthermore, they have obtained additional financing for the system, from a group of foreign oil companies whose workers will utilize the buildings. The oil company financing is being provided in exchange for the credits, but the oil companies are primarily interested in the project as a means of boosting the domestic oil market (the hope is that other Azeri cities will also opt for oil-fired district heating).

PROJECT ADDITIONALITY: The project is a free rider. Neither the local Azeri builders nor the foreign oil companies are pursuing the project primarily to obtain the credits; rather the builders want to ensure the most secure heat delivery system possible, while the oil companies are seeking to develop a new market for their product.

PROJECT EMISSIONS: When completed, the district heating system is expected to operate at a thermal efficiency of approximately 50 percent; i.e., 50 percent of the energy in the fuel consumed will be delivered in the form of heat to the hotel rooms and apartments. Therefore, given an emission factor of 173.9 lbs CO_2/mmBtu, the emissions rate (ER) for the project can be estimated as

$$ER = (173.9 \text{ lbs } CO_2/\text{mmBtu burned})/(0.5 \text{ mmBtus used/mmBtu burned})$$

$$ER = 347.8 \text{ lbs } CO_2/\text{mmBtu needed}$$

PROJECT BENCHMARKS: No district heating systems have been built in Azerbaijan in recent years. We will assume, for illustrative purposes, that a total of four such systems have been built in the FSU and Eastern Europe in the past five years. These systems, although new, have already experienced significant declines in efficiency due to the difficulty of obtaining spare parts to maintain such systems in the FSU. The key (fictional) data for these four systems are as follows:

Location	Fuel Used	Total Energy Delivered (mmBtus)	Efficiency* (Percent)	Emissions Rate* (lbs CO_2/mmBtu)
Moscow	Natural Gas	1.9	40	292.7
St. Petersburg	Oil	1.0	35	496.9
Kiev	Natural Gas	1.6	45	260.2
Almaty	Coal	1.2	45	456.2
Total/Average		5.7	41 (41.6)	376.5 (353.8)

*Averages shown are arithmetic. Weighted averages are in parentheses.

PROJECT ANALYSIS TABLE: Project Number IS1

	U.S. Proposal	EU Proposal	Full Technology Matrix	Hybrid Technology Matrix
Does project qualify?	For industrial practices, the eligibility threshold is set at X percentile of the efficiencies for facilities in the reference scenario. If $X>25^{th}$ percentile, then the project will qualify (Efficiency > 45%). Data are insufficient to support a value of $X<25^{th}$ percentile.	This project falls under the category of energy efficiency on the positive list, as it represents a significant improvement in existing energy production, and a significant improvement in energy transmission. However, because the terminology "significant" is insufficiently defined, it cannot be determined whether or not this technology meets the additionality criteria.	This project does not employ advanced, non-commercial technology, so it will not qualify under the full technology matrix. The project developer would be given the opportunity to prove additionality using the project-specific approach, but given that the project is a free rider, it is unlikely that additionality could be proved.	This project does not qualify as additional under the hybrid technology matrix.
Is the project correctly identified as either a free rider or an additional project?	No, if $X>25^{th}$ percentile. Indeterminate if $X<25^{th}$ percentile, due to the need for further data.	Indeterminate. Positive list terminology must be more clearly defined before making a positive determination.	Yes.	Yes.
Number of credits Awarded	For industrial practices, the baseline is determined as the weighted-average efficiency for facilities in the reference scenario, divided into the emissions factor for the fuel used by the project (oil). If $X>25^{th}$ percentile, then credits = 71.2 lbs CO_2/mmBtu (i.e., 418* lbs CO_2/mmBtu - 347.8 lbs CO_2/mmBtu)	Not applicable.	Because this project does not qualify under the technology matrix, it does not qualify for credits (i.e., 0 credits would be awarded).	Because this project does not qualify under the hybrid technology matrix, it does not qualify for credits (i.e., 0 credits would be awarded).
Error in credits Awarded	If $X>25^{th}$ percentile, then the error in credits awarded = 100 percent of the credits awarded (76.3 lbs CO_2/kWh)	Not applicable.	Project is a free rider, so the error in credits awarded is 0.	Project is a free rider, so the error in credits awarded is 0.

*418 lbs CO_2/mmBtu = (173.9 lbs CO_2/mmBtu burned)/(0.416 mmBtus/mmBtu burned)

METHODOLOGY ASSESSMENT: This non-additional, free rider project fails to be identified as such by both the U.S. and EU approaches. Under the U.S. approach, data is insufficient to determine a percentile threshold lower than 25 percent. Only four data points are given because no district heating systems have been built in Azerbaijan in recent years, and comparative data is lacking and/or difficult to collect from the FSU and Eastern Europe. For industrial practices, the eligibility threshold under the U.S. proposal is set at either the [X] percentile of the highest efficiency or lowest emissions rate for facilities in the reference scenario. With only four data points with which to make this determination, efficiencies were compared rather than emissions rates in order to attain a more accurate result, as the four plants varied greatly in fuel type.

This particular example illustrates that the U.S. approach has the potential to result in an erroneous determination of additionality if data is limited. Certainly, the ability of the percentile threshold test to yield an accurate determination of additionality is more likely when a more expansive range of data on similar and recent technologies is available. In this case, only four data points were available, making a percentile distribution test for additionality more likely to yield an inaccurate result. The problems resulting from the lack of data are compounded by maintenance problems in the FSU. Due to the difficulties of obtaining spare parts throughout much of the FSU, the four recent and comparable facilities included in the reference scenario have already experienced significant efficiency reductions. The project may compare favorably with the reference scenario facilities simply because it is newer, not necessarily better.

On the other hand, the technology matrix approaches succeeded in failing to qualify this free rider as additional. Despite the lack of data in this instance, the technology matrix employs tests that are not as dependent on an extensive data set in order to determine additionality, resulting in an accurate determination of non-additionality.

This free-rider project also fails to be identified as such under the EU approach. In fact, it is impossible to absolutely determine whether this project would qualify for credits under the EU's positive list. This project may fall under the category of energy efficiency, as it may represent a significant improvement in existing energy production and/or significant improvements in energy transmission. However, it is impossible to completely label this project as additional until the term "significant" in the context of the positive list is fully defined. One individual's concept of what constitutes a "significant" improvement may be entirely different from yet another individual's concept. The terminology under the positive list should be more fully defined and detailed in order to avoid future errors in additionality determinations under the EU approach.

PROJECT NUMBER: IS2

COUNTRY: Kazakhstan

SECTOR: Industrial

PROJECT TITLE: Cogeneration at Food Processing Plant

PROJECT DESCRIPTION: This project involves the construction of a new 5-MW, coal-fired Combined Heat and Power (CHP) plant at a food processing facility in Kazakhstan. The plant will provide both steam and electricity for on-site use only. Some sort of steam boiler plant is required on site to supply steam to the food processing plant. The food processing company decided to build a cogeneration plant rather than a simple steam production plant in order to take advantage of the high efficiencies inherent in cogeneration, and to provide their facility with a reliable source of electricity.

The Kazak food processor is supplying most of the financing required for building the CHP plant. However, a limited amount of additional funding is being provided by an American company, in exchange for the credits generated by the project.

PROJECT ADDITIONALITY: Cogeneration is used extensively in Kazakhstan and throughout the FSU, at a variety of industrial facilities. It is used because it can provide both steam and electricity to meet the needs of these facilities, in a highly efficient manner. This project is no different. The CHP plant is required to meet the needs of the on-site industrial processing plant; it is not being built to reduce emissions or gain credits. Therefore, the project is a non-additional free rider.

PROJECT EMISSIONS: The CHP plant has a design efficiency of 65 percent. Therefore, using the standard emission factor for bituminous coal of 205.3 lbs CO_2/mmBtu, the project's emission rate (ER) per mmBtu of total energy (electricity plus steam) output can be estimated as follows:

ER = (205.3 lbs CO_2/mmBtu)/(0.65 mmBtus output/mmBtu input)

ER = 315.8 lbs CO_2/mmBtu

PROJECT BENCHMARKS: We will assume that only two other cogeneration facilities have been built in Kazakhstan in recent years. However, if the geographic level of aggregation is expanded to include the FSU as a whole, a total of 15 new CHP plants have been built, to support new manufacturing facilities built to meet new domestic consumer markets. The following is the relevant (fictional) data for these 15 CHP plants.

86

Plant Name	Country	Fuel Used	Energy Output (mmBtus)	Efficiency* (Percent)	Emissions Rate* (lbs CO_2/mmBtu)
CHP1	Russia	Coal	250,000	61	336.6
CHP 2	Russia	Natural Gas	240,000	61	192.0
CHP 3	Russia	Natural Gas	260,000	63	185.9
CHP 4	Russia	Natural Gas	225,000	62	188.9
CHP 5	Russia	Natural Gas	245,000	60	195.2
CHP 6	Russia	Natural Gas	255,000	59	198.5
CHP 7	Russia	Natural Gas	254,000	57	205.4
CHP 8	Russia	Natural Gas	246,000	65	180.2
CHP 9	Russia	Natural Gas	247,000	64	183.0
CHP 10	Russia	Natural Gas	252,000	58	201.9
CHP A	Ukraine	Natural Gas	250,000	57	205.4
CHP B	Ukraine	Natural Gas	248,000	61	192.0
CHP C	Ukraine	Natural Gas	254,000	62	188.9
Kz 1	Kazakhstan	Coal	260,000	59	348.0
Kz 2	Kazakhstan	Coal	258,000	63	325.9
Az 1	Azerbaijan	Oil	240,000	59	294.7
Az 2	Azerbaijan	Oil	245,000	58	299.8
U 1	Uzbekistan	Oil	270,000	62	280.5
Ar 1	Armenia	Oil	248,000	64	271.7
Ar 2	Armenia	Oil	240,000	65	267.5
Coal Average	NA	Coal	256,000	61.0 (61.0)	336.8 (336.9)
Oil Average	NA	Oil	248,600	61.6 (61.6)	282.8 (282.8)
Gas Average	NA	Gas	248,000	60.8 (60.7)	193.1 (193.2)
Overall Average	NA	NA	249,000	61.0 (61.0)	237.1 (237.6)

*Averages shown are arithmetic. Weighted averages are in parentheses.

PROJECT ANALYSIS TABLE: Project Number IS2

	U.S. Proposal	EU Proposal	Full Technology Matrix	Hybrid Technology Matrix
Does project qualify?	If $X < 5^{th}$ percentile (out of the efficiencies for the 20 co-generation plants), then the project will not qualify. If $X > 10^{th}$ percentile, then the project will qualify (Efficiency > 64%).	The EU's positive list allows "advanced technologies" to qualify. However, it is unclear whether this means that all co-generators qualify because co-generation is itself considered an advanced technology, or whether only co-generators using "advanced technologies" will qualify.	Cogeneration is used extensively in the FSU; therefore, this project will not qualify as additional. The project developer would be given the opportunity to prove additionality using the project-specific approach, but given that the project is a free rider, it is unlikely that additionality could be proved.	This project does not qualify as additional under the hybrid technology matrix.
Is the project correctly identified as either a free rider or an additional project?	Yes, if $X > 10^{th}$ percentile. No, if $X < 5^{th}$ percentile.	Indeterminate. Positive list concepts and terms must be more clearly defined.	Yes.	Yes.
Number of credits Awarded	If $X > 10^{th}$ percentile, then the baseline is determined using the weighted average efficiency for the facilities multiplied by the emissions factor for the fuel used by the project (coal). Thus, credits would be awarded at a rate of 6.2 lbs CO_2/mmBtu (i.e., $(1/.64 - 1/.65) \times 205.3$ lbs CO_2/mmBtu).	Not applicable.	Because this project does not qualify under the technology matrix, it does not qualify for credits (i.e., 0 credits would be awarded).	Because this project does not qualify under the hybrid technology matrix, it does not qualify for credits (i.e., 0 credits would be awarded).
Error in credits Awarded	If $X > 10^{th}$ percentile, then error is equal to 100 percent of credits awarded (6.2 lbs CO_2/mmBtu).	Not applicable.	Project is a free rider, so the error in credits awarded is 0.	Project is a free rider, so the error in credits awarded is 0.

METHODOLOGY ASSESSMENT: This non-additional, free rider project fails to be identified as such under the EU's positive list and under certain circumstances under the U.S. approach. This is yet another example of the inadequate definition of concepts and terms under the EU's positive list. This project involves the construction of a co-generation plant at a food processing facility, which will provide both steam and electricity. Co-generation, in and of itself, is often considered an advanced technology. The EU's positive list allows "advanced technologies" for co-generation plants to qualify as additional under the category of energy efficiency. However, it is not clear whether this means that all co-generators qualify because co-generation is considered an advanced technology, or whether only co-generators equipped with advanced technologies would qualify.

Under the U.S. approach, if X is greater than the 10th percentile (out of the efficiencies of the twenty co-generation plants), then the project will qualify as additional (i.e., X>64%). However, if X is less than the 5th percentile, then the project fails to qualify as additional. As discussed previously, by its very nature, the U.S. percentile threshold test will qualify a certain percentage of business-as-usual projects, depending on the value that is chosen for "X." In this case, the final qualification test will be either accurate or erroneous depending on that choice. If the choice is made to set "X" at the most stringent level, then the project will be correctly identified as a free rider.

PROJECT NUMBER: IS3

COUNTRY: Argentina

SECTOR: Industrial

PROJECT TITLE: Variable Frequency Drives

PROJECT DESCRIPTION: This project involves retrofits and improvements to the motor/motor drive systems at an industrial plating plant in Argentina. The affected motors run the plant's ventilation system. In the past, they were operated at full load on a 24/7 operating schedule, even though the process areas ventilated via the motors are at times not utilized. Airflow was controlled via dampers.

The project involves the retrofitting of variable frequency drives (VFDs) to the fan motors. In addition, a direct digital control system (DDCS) was added to control fan speed, based on process area utilization.

The industrial plating plant is a joint venture between an U.S. company and an Argentine company. The partners plan to share any credits generated by the project.

PROJECT ADDITIONALITY: Because the industrial plating plant operates on a 24/7 schedule, the energy savings generated by the project are limited. Basically, the VFDs and DDCS allow a slight reduction in the load on the motors at all times, and a complete cessation of motor use during shift changes. The resulting total reduction in energy used by the fan motors is 10 percent. Because this reduction is fairly small, the project proved to be sub-economic when evaluated without the credits. However, when the value of the credits expected by the project was factored in, the project met the partner's economic feasibility requirements. Therefore, the project is not a free rider.

PROJECT EMISSIONS: Before the project, the affected motors utilized 200 kW of power, on average. The emissions factor for power supplied to the plant has been estimated at 1.2 lbs CO_2/kWh. Therefore, taking into account the fact that the project reduces the load on the motors by 10 percent, the project emissions rate (ER) can be computed as follows:

$$ER = (0.9)(200 \text{ kW})(1.2 \text{ lbs } CO_2/\text{kWh})$$

$$ER = 216 \text{ lbs } CO_2/\text{h}$$

PROJECT BENCHMARKS: There is no available data on the utilization of industrial fan motors, either for Argentina or surrounding countries. However, a recent market research study indicates that of the 1000 industrial motor/motor drive systems sold in Argentina in the past 5 years, 10 percent were equipped with VFDs.

PROJECT ANALYSIS TABLE: Project Number IS3

	U.S. Proposal	EU Proposal	Full Technology Matrix	Hybrid Technology Matrix
Does project qualify?	The data required to establish a percentile threshold (i.e., efficiency or emissions rate) is unavailable.	This project qualifies under the EU's positive list. Specifically, it falls under the category of demand side management. This is an energy conservation project, resulting in improvements in industrial energy consumption. Therefore, it qualifies as additional under the EU's positive list.	This project does not qualify as additional under the full technology matrix approach. This is a conventional technology that has penetrated the market. Project developers may still attempt to prove additionality using the project-specific approach.	The project will not qualify as additional under the technology matrix.
Is the project correctly identified as either a free rider or an additional project?	Indeterminate. Additionality test does not account for projects that lack efficiency or emissions rate data.	Yes.	No. Because the project is small, the fallback of the project-specific approach is unlikely to be utilized due to the high transaction costs associated with this approach.	No. Because the project is small, the fallback of the project-specific approach is unlikely to be utilized due to the high transaction costs associated with this approach.
Number of credits Awarded	Indeterminate.	Not applicable.	The project did not qualify as additional, therefore, the project is not awarded any credits (i.e., credits awarded = 0).	The project did not qualify as additional, therefore, the project is not awarded any credits (i.e., credits awarded = 0).
Error in credits Awarded	Unknown.	Not applicable.	Because this project is awarded no credits, but needs the incentive of credits in order to be economically feasible, it will not be undertaken.	Because this project is awarded no credits, but needs the incentive of credits in order to be economically feasible, it will not be undertaken.

METHODOLOGY ASSESSMENT: This project failed to be accurately classified as additional under all approaches except for the EU approach. Under the EU approach, this energy conservation project clearly falls under the category of demand side management, as it employs technology that results in improvements in industrial energy consumption. Under the U.S. approach, the data required to establish a percentile threshold test are unavailable; therefore, additionality as well as the number of credits awarded cannot be determined. For industrial practices, a percentile threshold test must be established using either emissions rates or efficiencies within a reference scenario. In this case, because this is an energy conservation project that employs technology to allow for the adjustment of fan speed, emissions rates and efficiencies are irrelevant. This is an energy utilization project as opposed to an energy efficiency project. Thus, it would be extremely difficult to establish a percentile threshold in any meaningful way due to the nature of the project. In this particular example, there exists a lack of comparison points for an accurate additionality determination to be made under the U.S. approach. However, the basic predicament in this example is not a lack of data. The problem rests in the fact that the test that has been established to determine additionality under the U.S. approach fails to account for industrial practice projects such as this, that do not improve efficiency per se. The U.S. should perhaps establish a back-up test to account for industrial practice project scenarios that fall outside of the "efficiency and/or emissions rate" box.

Although the project fails to qualify under the full and hybrid technology matrix approaches, the project developers could attempt to demonstrate project additionality using the project-specific approach. However, when a project is small, such as this one, the transaction costs may be too high to warrant use of the project-specific approach, thus denying some truly legitimately smaller additional projects from qualifying for credits. Although the technology matrix includes the fallback of the project-specific approach, this fallback is not a panacea, and it may not be an option for developers of small projects.

The implications of the failure of this project to qualify for credits are that the project would not be undertaken, and the fan motors would continue to operate at full load on a 24/7 schedule. The VFD and DDCS technologies are truly additional, and as such, require the incentive of credits in order to be deemed feasible. The failure to qualify truly additional projects such as this one will reduce the amount of credits awarded to the U.S. and other developed countries, thereby raising the costs of implementation.

PROJECT NUMBER: IS4

COUNTRY: Brazil

SECTOR: Industrial

PROJECT TITLE: Retrofit of Energy Efficient Motors

PROJECT DESCRIPTION: A Brazilian lock canal, built in the 1960s, uses a number of large electric motors to operate the locks. For the first time since they were originally installed, the canal operator plans to replace the original motors with new, custom-built, highly efficient motors. The resulting savings in electricity costs are expected to payback the costs of the motors in 2 years. Although most of the required capital will be provided by the canal operator, limited additional financing will be supplied by an American company, in exchange for the credits.

PROJECT ADDITIONALITY: The new motors, although custom-built to very exacting specifications, do not represent a new technology. The project is being undertaken because it has a short payback period (2 years) with or without the credits; therefore the project is a free rider.

PROJECT EMISSIONS: The original motors required a total of 20 MWs of power when in operation, and they operated, on average, 15 percent of the year. The power is supplied by an on-site gas-fired gas turbine, with a heat rate of 15 mmBtus/MWh. The project is expected to improve motor efficiency by 7 percent. Therefore, given an emissions factor for gas of 1117.1 lbs CO_2/mmBtu, the estimated emissions rate (ER) for the project can be computed as follows:

$$ER = (0.93)(20 \text{ MW})(15 \text{ mmBtus/MWh})(117.1 \text{ lbs } CO_2/\text{mmBtu})(0.15 \text{ h/h})$$

$$ER = 4900 \text{ lbs } CO_2/\text{h, or } 2.45 \text{ tons } CO_2/\text{h}$$

PROJECT BENCHMARKS: Based on motor manufacturers' data, the lock motors are more efficient than 99 percent of all models manufactured in the western hemisphere in the last 5 years.

PROJECT ANALYSIS TABLE: Project Number IS4

	U.S. Proposal	EU Proposal	Full Technology Matrix	Hybrid Technology Matrix
Does project qualify?	If $X > 1^{st}$ percentile, then the project will qualify (efficiency of project is greater than 99 percent of all models manufactured in the western hemisphere in the last 5 years).	The project falls under two positive list categories: (1) energy efficiency (significant improvements in industrial processes and energy transmission), and (2) demand side management (improvement in industrial energy consumption). The project qualifies under the second category, but it is unclear whether project qualifies under the first category, because definition of "significant" is unclear.	The project does not involve advanced, non-commercial technology. Therefore, it will not qualify under the technology matrix. Project developers may attempt to prove additionality using the project-specific approach, but given that the project is being undertaken because it has a short payback period, with or without credits, it is unlikely to qualify.	The project will not qualify as additional under the technology matrix, and is unlikely to qualify as additional under the project-specific approach.
Is the project correctly identified as either a free rider or an additional project?	No. The comparison of project to an average of recent plants will not always yield and accurate result.	Indeterminate. Positive list terminology and category distinctions must be more clearly defined.	Yes.	Yes.
Number of credits Awarded	It is unclear whether the credits for a retrofit project are computed using a benchmark or the efficiency of old motors prior to replacement. If the former, data is unavailable. If the latter, credits = 370 lbs CO_2/h. (i.e., 5,270 lbs CO_2/h* (ER new motors) - 4,900 (ER old motors)).	Not applicable.	Project does not qualify for any credits.	Project does not qualify for any credits.
Error in credits Awarded	Project is a free rider; thus, error is equal to 100 percent of the credits awarded.	Not applicable.	Project is correctly identified as a free rider; thus, error in the credits awarded is zero.	Project is correctly identified as a free rider; thus, error in the credits awarded is zero.

*Emission rate for old motors determined using the following formula: $(1.00)(20 \text{ MW})(15 \text{ mmBtus/mWh})(117.1 \text{ lbs } CO_2/\text{mmBtu})(0.15 \text{ h/h}) = 5,270 \text{ lbs } CO_2/\text{h}$

94

METHODOLOGY ASSESSMENT: This non-additional, free rider project fails to be identified as such under both the U.S. and the EU approaches. Under the U.S. approach, the percentile distribution test reveals that the efficiencies of the project's lock motors are greater than 99 percent of all models manufactured in the Western Hemisphere in the last five years. Thus, the project qualifies as additional. This project demonstrates a deep-seated problem with the U.S. proposal's percentile distribution test. That is, a certain number of non-additional or free rider projects will qualify for credits, even if, as in this case, the percentile test is set at the most stringent level. By simply requiring emission reduction projects to be "significantly better than the average," the U.S. percentile test guarantees that a certain number of business-as-usual projects will qualify for credits. Because the project is not identified as a free rider, the error in credits awarded is equal to 100 percent (370 lbs CO_2/h). In contrast, the technology matrix approach analyzes the technology itself based on market penetration and economic feasibility. These economic criteria directly address the issue underlying free ridership; namely, would the project be undertaken absent the economic incentive of credits. The comparison of the project emission rate with a benchmark value, under the U.S. threshold test, does not directly address the issue of free ridership. It is at best only a proxy test, and hence, may fail even when "X" is set at a very stringent level. In this case, the technology matrix and hybrid technology matrix approaches succeeded in correctly identifying the project as a free rider. Thus, the project failed to gain credits under these approaches.

Under the EU approach, whether or not the project will qualify is indeterminate. This particular project may fall under two discrete categories under the EU's positive list: (1) energy efficiency, in the sub-category of significant improvements in industrial processes and energy transmission, and (2) demand side management (improvement in industrial energy consumption). The project may indeed qualify as additional under the second category, as the motor is an end use device. However, it is unknown whether or not it will definitely qualify as additional under the first category due to the lack of a definition of the term "significant." As previously discussed, the terminology in this first category is too vague to reach a conclusive determination of additionality. The fact that a single project could potentially fall into two separate categories under the EU positive list, resulting in different determinations of additionality is extremely problematic. It is imperative that categories included in the list be exclusive and clearly defined. Furthermore, the criteria used within each category to determine whether a project qualifies must be objective and clearly defined. As the positive list is currently written, it is impossible to determine in any objective manner whether a given project meets the criteria of "significant improvement."

It is important to note that not only does the U.S. approach fail to screen out this free rider project, but also further errors will be generated by this method during baseline development. It is unclear, based on the existing documentation, whether the credits for this industrial retrofit project are to be computed using a benchmark or the actual efficiency of the old motors prior to the replacement. The number of credits awarded using the latter approach yielded an error in credits awarded of 370 lbs CO_2/h, while data was unavailable to compute the number of credits awarded using a benchmark. The technology matrix approach makes a clear distinction between new facility projects and projects involving retrofits to existing facilities. Only the former projects use benchmarks. The number of credits awarded for a retrofit of advanced qualifying technologies to an existing facility is determined using the project-specific approach. Thus, the baseline for a retrofit project must be computed using historical data for the affected facility. The U.S. method, as previously noted, should be either modified, or better explained, to make or clarify this same fundamental distinction between new facility and retrofit projects.

PROJECT NUMBER: IS5

COUNTRY: China

SECTOR: Industrial

PROJECT TITLE: Coke Oven Underfiring Rate Improvement

PROJECT DESCRIPTION: This project involves the retrofitting of new equipment to existing coke oven batteries at a Chinese steel mill. The new equipment incorporates advanced firing technology designed to improve the underfiring rate, thereby reducing the energy used per ton of coal charged in the ovens.

The steel mill is a joint venture between a U.S. company and a Chinese company. Any credits generated by the project will be distributed to each partner according to its equity share in the joint venture.

PROJECT ADDITIONALITY: The project introduces new, advanced coke oven firing technology into China for the first time. Although the energy and cost reduction benefits of the project were important factors in the partners' decision making, it was the prospect of obtaining credits that tipped the balance in favor of proceeding with the project despite the inherent risks in the use of new technology.

PROJECT EMISSIONS: The project will reduce the amount of coal consumed per ton of coal charged in the ovens, from 2.5 mmBtus to 2.3 mmBtus. Thus given the coal emission factor of 205.3 lbs CO_2/mmBtu, the project's emission rate can be computed as follows:

$$ER = (2.3 \text{ mmBtus/ton})(205.3 \text{ lbs } CO_2/\text{mmBtu}) = 472.2 \text{ lbs } CO_2/\text{ton charged}$$

PROJECT BENCHMARKS: We will assume that five new coke oven batteries have been opened in China in the past 5 years. All of these coke ovens utilize coal as their primary energy source. The relevant (fictional) data for these batteries are as follows:

Unit ID	Coal Charge (tons)	Average Energy Usage* (mmBtus/ton charged)	Emissions Rate* (lbs CO_2/ton charged)
No. 1	200,000	2.5	513.2
No. 2	185,000	2.6	533.8
No. 3	190,000	2.8	574.8
No. 4	210,000	2.4	492.7
No. 5	225,000	2.5	513.2
Total/Average	1,010,000	2.56 (2.55)	525.5 (524.3)

*Averages shown are arithmetic. Weighted averages are in parentheses.

PROJECT ANALYSIS TABLE: Project Number IS5

	U.S. Proposal	EU Proposal	Full Technology Matrix	Hybrid Technology Matrix
Does project qualify?	If X>20th percentile for energy usage, then the project will qualify (energy usage < 2.4 mmBtus/ton charged). If X<20th percentile, then more data is needed to develop an adequate percentile distribution.	This project qualifies as additional under the EU's positive list. Specifically, under the energy efficiency category, it represents an *advanced technology* leading to significant improvements in industrial processes.	The project involves an advanced, non-commercial technology being introduced into China for the first time. Projects using this technology will therefore automatically qualify as additional under the technology matrix.	Project will automatically qualify as additional.
Is the project correctly identified as either a free rider or an additional project?	Yes, if X>20th percentile.	Yes.	Yes.	Yes.
Number of credits Awarded	It is unclear whether the credits for a retrofit project would be computed using a benchmark or the actual emissions rate prior to the retrofit. In the former case, for industrial practices, the baseline is determined as the weighted-average emission rate per unit of output. If X>20th percentile, then credits would be awarded at a rate of 52.1 lbs CO_2/ton charged (i.e., 524.3 - 472.2 lbs CO_2/ton charged). In the latter case, credits would be awarded at a rate of 41.1 lbs CO_2/ton charged (i.e., 513.3* - 472.2).	Not applicable.	Because this project involves the retrofit of advanced qualifying technologies to an existing facility, the project specific approach is used to estimate the baseline. Thus, credits are determined by subtracting the ER of the project from the ER of the plant prior to the retrofit. credits will thus be awarded at a rate of 41.1 lbs CO_2/ton charged (i.e., the ER with the project--472.2--subtracted from the ER before the project--513.3*).	Under the Hybrid Technology Matrix, estimated credits would be determined using the U.S. approach for determination of credits. Therefore, credits would either be awarded at a rate of 52.1 lbs CO_2/ton charged (i.e., 524.3 - 472.2 lbs CO_2/ton charged), or at a rate of 41.1 lbs CO_2/ton charged (i.e., 513.3* - 472.2).
Error in credits Awarded	Unknown.	Not applicable.	Unknown.	Unknown.

*The ER prior to the project was determined using the following equation: (2.5 mmBtus/ton)(205.3 lbs CO_2/mmBtu) = 513.3 lbs CO_2/ton charged.

99

METHODOLOGY ASSESSMENT: This project was correctly identified as additional by all approaches. However, under the U.S. approach, while the project was identified as additional only if "X" > 20th percentile for energy usage, if "X" < 20th percentile, further data is required to develop an adequate percentile distribution. Again, this example demonstrates the relatively substantial data requirements of the U.S. approach.

Under the U.S. approach, errors will be generated during baseline development. It is unclear, based on the existing documentation, whether or not the baseline for a retrofit project is to be computed using the actual emissions rate of the project prior to the retrofit, or a sector benchmark. In this particular case, the number of credits awarded is greater (i.e., 52.1 lbs CO_2/ton charged) when the sector benchmark is used to determine the number of credits awarded rather than the actual emissions rate of the project prior to the retrofit (i.e., 41.1 lbs CO_2/ton charged).

The technology matrix approach makes a clear distinction between new facility projects and projects involving the retrofit of advanced qualifying technologies to an existing facility. In the latter case, the project-specific approach is used to estimate the baseline; thus, historical data for the affected facility is used to compute the baseline. As stated previously, the U.S. approach should be either modified, or better explained, to make or clarify this same fundamental distinction between new facility and retrofit projects.

PROJECT NUMBER: IS6

COUNTRY: Tajikistan

SECTOR: Industrial

PROJECT TITLE: PFC Reductions at Aluminum Plant

PROJECT DESCRIPTION: Emissions of perfluorocarbons (PFCs)—specifically perfluoromethane (CF_4) and perfluoroethane (C_2F_6)—occur during the electrochemical aluminum smelting process. Specifically, PFCs are emitted during discrete periods of process inefficiencies known as anode effects. Anode effects occur when the amount of alumina in solution inside the reduction cell drops below the level necessary to drive the desired chemical reaction. When this occurs, the voltage across the cell increases rapidly, the chemical reaction reverses, such that previously reduced aluminum is re-oxidized into alumina ore, and fluoride is emitted in the form of PFCs. PFCs are highly potent greenhouse gases with global warming potentials (GWPs) several thousand times greater than that of carbon dioxide.

The frequency and duration of anode effects can be reduced through computerized process control systems that closely monitor and adjust the amount of alumina in solution. This project involves the installation and optimization of just such a system in Tajikistan's sole existing aluminum smelter. Because the project is expected to yield a large quantity of credits (in carbon equivalent tons) for a relatively limited investment, an American aluminum company has agreed to provide full project financing in exchange for all credits generated. The American company has also agreed to provide needed technical expertise in the installation, optimization, and operation of the system, for a period of up to 1 year.

PROJECT ADDITIONALITY: Although Tajikistan's smelter will be the first in the region to install the new, advanced process control technology, it is now being used extensively in the United States and other developed countries. However, although the smelter could perhaps finance the project on its own, and would do so to gain the cost benefits of the resulting process efficiency improvements, it lacks the technical expertise required to install and operate the new system. In short, without the American company's pledge of technical support, the project cannot go forward, and that pledge is being made solely for the purpose of gaining credits. The award of credits is a prerequisite to project implementation, and the project is therefore not a free rider.

PROJECT EMISSIONS: The smelter currently emits an average of 0.0005 tons of CF_4 and 0.00005 tons of C_2F_6, per ton of aluminum produced. The project is expected to reduced emissions of both gases by 40 percent; therefore the project's emission rates (ERs) can be computed as follows:

$$ER \ (CF_4) = (0.6)(0.0005 \text{ tons } CF_4/\text{ton Al}) = 0.0003 \text{ tons } CF_4/\text{ton Al}$$

$$ER \ (C_2F_6) = (0.6)(0.00005 \text{ tons } C_2F_6) = 0.00003 \text{ tons } C_2F_6/\text{ton Al}$$

PROJECT BENCHMARKS: Although Tajikistan has only one aluminum smelter, we will assume that smelter data covering the wider region have been obtained. Since none of these smelters is using the new computerized process control system, their PFC emissions are relatively high compared to the project's emissions. The relevant (fictional) data are as follows:

Smelter	Aluminum Production (Tons)	CF$_4$ Emissions Rate* (Tons CF$_4$/Ton Al)	C$_2$F$_6$ Emissions Rate* (Tons C$_2$F$_6$/Ton Al)
No. 1	150,000	0.0005	0.00005
No. 2	2000,00	0.0006	0.00006
No. 3	250,000	0.0006	0.00006
No. 4	100,000	0.0005	0.00005
No. 5	90,000	0.0007	0.00007
No. 6	175,000	0.0006	0.00006
No. 7	150,000	0.0008	0.00008
No. 8	140,000	0.0007	0.00007
No. 9	190,000	0.0006	0.00006
No. 10	110,000	0.0005	0.00005
Total/Average	1,555,000	0.00061 (0.00061)	0.000061 (0.000061)

*Averages shown are arithmetic. Weighted averages are shown in parentheses.

PROJECT ANALYSIS TABLE: Project Number IS6

	U.S. Proposal	EU Proposal	Full Technology Matrix	Hybrid Technology Matrix
Does project qualify?	If $X > 10^{th}$ percentile, then the project will qualify as additional (ER for CF_4 < 0.0005 tons CF_4/ton Al and ER for C_2F_6 < 0.00005 tons C_2F_6/ton Al).	This project fails to qualify as additional under the EU's positive list. The positive list includes only energy-related projects that reduce carbon dioxide emissions.	The computerized process control system to be installed is an advanced, non-commercial technology in Tajikistan. Projects using this technology will automatically qualify as additional under the technology matrix.	Project will automatically qualify as additional.
Is the project correctly identified as either a free rider or an additional project?	Yes.	No. Project improves process efficiency, not energy efficiency; thus, positive list categories are irrelevant.	Yes.	Yes.
Number of credits Awarded	It is unclear whether the credits for a retrofit project would be computed using a benchmark or the actual ER prior to retrofit. If the former, credits are determined as weighted-averages of the emissions rates per unit of output, for comparable facilities. The estimated credits for CF_4 = 0.00061 - 0.0003 = 0.00031 tons CF_4/ton Al. The estimated credits for C_2F_6 = 0.000061 - 0.00003 = 0.000031 tons C_2F_6/ton Al. If the latter, credits for CF_4 = 0.0005 - 0.0003 = 0.0002 tons CF_4/ton Al. credits for C_2F_6 = 0.00005 - 0.00003 = 0.00002 tons C_2F_6/ton Al.	Not applicable.	Because this project involves the retrofit of advanced qualifying technologies (in this case, a computerized process control system) to an existing facility, the project specific approach is used to estimate the baseline. Thus, credits are determined by subtracting the ER of the project from the ER of the plant prior to the retrofit. Thus, the estimated credits for CF_4 = 0.0005 - 0.0003 = 0.0002 tons CF_4/ton Al. The estimated credits for C_2F_6 = 0.00005 - 0.00003 = 0.00002 tons C_2F_6/ton Al.	The values for credits awarded are determined using the U.S. proposal approach. Thus, the estimated credits would either be 0.00061 - 0.0003 = 0.00031 tons CF_4/ton Al and 0.000061 - 0.00003 = 0.000031 tons C_2F_6/ton Al using the benchmark to determine credits or 0.0005 - 0.0003 = 0.0002 tons CF_4/ton Al and 0.00005 - 0.00003 = 0.00002 tons C_2F_6/ton Al using the emissions rate prior to the retrofit to determine credits.
Error in credits Awarded	Unknown.	Not applicable.	Unknown.	Unknown.

METHODOLOGY ASSESSMENT: This additional project was correctly identified as such by all approaches, except for the EU approach. The project fails to qualify under the EU's positive list because the positive list includes only energy-related projects that reduce carbon dioxide emissions. In this particular project scenario, emissions of perfluorocarbons (PFCs) are reduced using a computerized process control system. Thus, the project improves process efficiency, not energy efficiency. As currently written, the EU's positive list automatically screens out all non energy-related, non CO_2-emission reducing projects. Many such projects are likely to prove to be highly cost effective means of reaching global emission reduction goals. The positive list should be expanded to include these significant emission reduction opportunities.

Under the U.S. approach, errors will be generated during baseline development. It is unclear, based on the existing documentation, whether or not the baseline for a retrofit project is to be computed using the actual emissions rate of the project prior to the retrofit, or a sector benchmark. In this particular case, the number of credits awarded is greater (i.e., 0.00031 tons CF_4/ton Al and 0.000031 tons C_2F_6/ton Al) when the sector benchmark is used to determine the number of credits awarded rather than the actual emissions rate of the project prior to the retrofit (i.e., 0.0002 tons CF_4/ton Al and 0.00002 tons C_2F_6/ton Al).

The technology matrix approach makes a clear distinction between new facility projects and projects involving the retrofit of advanced qualifying technologies to an existing facility. In the latter case, the project-specific approach is used to estimate the baseline; thus, historical data for the affected facility is used to compute the baseline. As stated previously, the U.S. approach should be either modified, or better explained, to make or clarify this same fundamental distinction between new facility and retrofit projects.

PROJECT NUMBER: IS7

COUNTRY: China

SECTOR: Industrial

PROJECT TITLE: Coal Ash Utilization

PROJECT DESCRIPTION: The cement manufacturing process is a significant source of carbon dioxide emissions. This process involves the heating (or calcination) of calcium carbonate (limestone) in a kiln, to produce lime and carbon dioxide. The lime is combined with other materials to produce clinker (an intermediate product in the manufacture of cement), while the carbon dioxide is emitted to the atmosphere.

The fly ash produced from the combustion of coal in power plant boilers is normally an unwanted pollutant. However, when captured this fly ash can be used to replace the calcium carbonate in cement, up to a level of approximately 25 percent. The resulting product actually possesses improved properties vis a vis pure lime-based cement for many applications, and carbon dioxide emissions are reduced in proportion to the amount of calcium carbonate replaced.

This project involves the use of coal fly ash at a cement plant in China. The fly ash will be provided to the cement plant by a U.S.-owned power plant, in exchange for the credits generated by the project. The fly ash will be combined with calcium carbonate in a 20/80 mixture.

PROJECT ADDITIONALITY: This project represents one of the first attempts to use fly ash in the manufacture of cement in China. The Chinese company that owns the cement plant was induced to undertake the project when offered the fly ash free of charge (in exchange for the credits); without this incentive, the risk was considered too great to warrant project implementation. In short, the project requires credits in order to be undertaken, and hence it is not a free rider.

PROJECT EMISSIONS: The cement plant emits 0.8 tons of CO_2/ton of cement. By reducing the amount of limestone that must be calcinated in the kilns by 20 percent, the project reduces the plant's emission rate (ER) to 0.64 tons CO_2 per ton of cement:

$$ER = (0.8)(0.8 \text{ tons } CO_2 \text{ per ton of cement}) = 0.64 \text{ tons } CO_2 \text{ per ton of cement}$$

PROJECT BENCHMARKS: The value of 0.8 tons of CO_2 per ton of cement is a standard emission factor for cement production, which can be assumed to provide a reasonably accurate emissions estimate for all cement plants that do not use fly ash as a replacement material. We assume that, in China, only 1 percent of the existing cement plants have begun using fly ash. Therefore, the project's emission rate can be assumed to be well below the emission rate of at least 99 percent of all existing Chinese cement plants.

PROJECT ANALYSIS TABLE: Project Number IS7

	U.S. Proposal	EU Proposal	Full Technology Matrix	Hybrid Technology Matrix
Does project qualify?	If "X" > the 1st percentile then the project will qualify (project's emissions rate is below 99 percent of existing Chinese cement plants)	This project fails to qualify as additional under the EU's positive list. The positive list includes only energy-related projects that reduce carbon dioxide emissions	The project does not involve an advanced non-commercial technology. Therefore, the project would not qualify. The project developers would have the opportunity to qualify the project under the project-specific approach.	The project does not involve an advanced non-commercial technology. Therefore, the project would not qualify. The project developers would have the opportunity to qualify the project under the project-specific approach.
Is the project correctly identified as either a free rider or an additional project?	Yes.	No. Project does not improve efficiency, and the positive list screens out all non-energy related, non CO_2-emissions reducing projects.	Possibly, if the project developers choose to utilize project specific.	Possibly, if the project developers choose to utilize project specific.
Number of credits Awarded	It is unclear whether the credits for a project such as this one, involving a modification to an existing facility, would be computed using a benchmark for comparable facilities or the emission rate of the cement plant prior to the project. However, in this case, the standard emission factor of 0.8 ton CO_2/ton of cement would apply to both the cement plant in question and 99 percent of all comparable facilities. Hence, either way, credits = 0.8 tons CO_2/tons cement – 0.64 tons CO_2/tons cement = 0.16 tons CO_2/tons cement	Not Applicable	Under project-specific credits= 0.8 tons CO_2/tons cement – 0.64 tons CO_2/tons cement = 0.16 tons CO_2/tons cement	Under project-specific credits= 0.8 tons CO_2/tons cement – 0.64 tons CO_2/tons cement = 0.16 tons CO_2/tons cement
Error in credits Awarded	Unknown	Not Applicable	Unknown	Unknown

METHODOLOGY ASSESSMENT: This additional project is correctly identified as such by the U.S. proposal and is incorrectly identified under the EU proposal. Under the technology matrix, the project would not qualify, but may qualify under the project specific approach.

Under the EU's positive list, the project fails to qualify because the list only includes energy-related projects that reduce carbon dioxide emissions. In this particular project scenario, carbon dioxide emissions are reduced by replacing some of the calcium carbonate in cement with fly ash. Thus, the project merely alters an industrial process rather than improving the energy efficiency of the process. As currently written, the positive list automatically screens out all non energy-related, non CO_2-emissions reducing projects. Many such projects are likely to prove to be highly cost effective means of reaching global emission reduction goals. The positive list should be expanded to include these significant emission reduction opportunities.

As for the technology matrix, this project underscores a key aspect of this approach, that it will reject any project that does not involve a new technology. Therefore, industrial sector projects utilizing new processes that result in true emissions reductions will always fail the technology matrix additionality test and will not qualify for credits. However, this does not mean that these projects will never be undertaken. Unlike other approaches and proposals, the technology matrix gives project developers the opportunity to fall back on the project-specific approach. In other words, although this coal ash project would fail to qualify for credits under the technology matrix, project developers would have an opportunity to qualify the project under the project-specific approach. In this case, given that the cement company would not take on the project without credit incentives, the project will likely qualify under the project-specific approach.

As for the U.S. proposal, this project shows that new industrial processes or at least processes that are not widely used in a host country, that result in true emission reductions will generally qualify for credits.

PROJECT NUMBER: IS8

COUNTRY: Chile

SECTOR: Industrial

PROJECT TITLE: Building Insulation Improvement

PROJECT DESCRIPTION: This project involves the installation of improved insulation at a 500,000 square-foot pharmaceutical plant in Concepcion, Chile (latitude 37^0 South). The building shell consists of concrete block. The building is 30 years old, and the original insulation, which was inadequate, has deteriorated. The project also involves fixing leaks along the ventilation ducts, windows, and doors. The building utilizes natural gas for heat; there is no air conditioning.

The facility produces a variety of prescription and non-prescription drugs for both the domestic and export markets. The facility is wholly owned by the Chilean subsidiary of a major U.S. pharmaceutical company. The subsidiary is providing limited project funding; however, the bulk of the financing is being supplied by the parent company in exchange for the project's credits.

PROJECT ADDITIONALITY: The project is part of an international emissions reduction program undertaken by the parent corporation. Under this program, the company announces annual goals for the number of credits it will obtain through its international emission reduction efforts. It then donates the credits to an environmental NGO, which has pledged to permanently remove the credits from the market. The international emissions reduction program is part of the company's high-priority effort to maintain an image of good corporate citizenship and environmental stewardship.

Although the project is economically viable absent the credits, it was never recognized as a possible project by the Chilean subsidiary. The parent company identified it as a possible project during the audit it performed of its various foreign subsidiaries; the audit was part of the international emissions reduction program. The audit was designed to identify projects that are both economically viable and capable of generating significant quantities of credits. The parent company's goal in undertaking the project is to obtain the credits for application towards its annual credit goals, in a cost-effective manner. Had the parent company not undertaken its international program to gather credits, the audit would not have been undertaken, and the possibility of implementing the project would have remained undiscovered. The project is therefore not a free rider.

It should be noted that, in order to meet its credit goals, the parent corporation is prepared to incur whatever transaction costs are necessary to qualify the project.

PROJECT EMISSIONS: The space heater utilized, on average, 50,000 Btus per square foot of floor space per year prior to the project, at the annual average temperature range in Concepcion of 50 to 60 degrees F. Based on the energy audit, it is expected that the building's heat consumption will be reduced by approximately 20 percent once the project is completed. Hence, using the emissions factor for natural gas of 117.1 lbs CO_2/mmBtu, the project emissions rate (ER) can be computed as follows:

$$ER = (0.8)(0.05 \text{ mmBtus/ft}^2\text{-yr})(117.1 \text{ lbs } CO_2/\text{mmBtu}) = 4.7 \text{ lbs } CO_2/\text{ ft}^2\text{-yr}$$

PROJECT BENCHMARKS: We will assume that there is only a limited amount of data available on industrial buildings that are new, or that have newly-installed insulation. The available data are as follows:

Facility Name	Location (Latitude)	Type of Facility	Building Type	Energy Used for Heating (Btus/ft²-yr)	Emissions Rate (lbs CO₂/ ft²-yr)
Acme Mill	33^0 S	Steel Mill	Corrugated iron	20,000	4.1
ABC Chemicals	33^0 S	Chemical	Concrete block	80,000	13.9
XYZ Mill	42^0 S	Textile Mill	Concrete block	100,000	11.7
AAA Mill	18^0 S	Textile Mill	Brick	5,000	0.6
Average				51,250	7.6

PROJECT ANALYSIS TABLE: Project Number IS8

	U.S. Proposal	EU Proposal	Full Technology Matrix	Hybrid Technology Matrix
Does project qualify?	If "X" > the 25th percentile then the threshold would be at least 0.6 lbs CO_2/ ft^2-yr. The project emissions rate of 4.7 lbs CO_2/ ft^2-yr falls above that threshold and therefore would not qualify.	Project falls under two positive list categories: (1) energy efficiency, (significant improvements in buildings), and (2) demand side management (improvements in commercial energy consumption). The project qualifies under the second category. It is unclear whether the project qualifies under the first category, because the definition of "significant" is unclear.	The project does not involve an advance non-commercial technology; therefore, the project would not qualify. The project developers would have the opportunity to qualify the project under the project specific approach.	The project does not use an advance non-commercial technology but may qualify under the market penetration test.
Is the project correctly identified as either a free rider or an additional project?	No. Data set facility may not truly be "comparable" to the project facility, both in geography and energy usage. U.S. needs to better clarify the terms "recent and comparable" for the percentile threshold test.	Indeterminate. Positive list terminology and category distinctions must be more clearly defined.	Possibly, if the project developers choose to utilize project-specific approach.	Possibly, if the project developers choose to utilize project-specific approach.
Number of credits Awarded	Project does not qualify for credits	Not Applicable	credits = 5.9 lbs CO_2/ ft^2-yr – 4.7 lbs CO_2/ ft^2-y = 1.2 lbs CO_2/ ft^2-y	credits = 5.9 lbs CO_2/ ft^2-yr – 4.7 lbs CO_2/ ft^2-y = 1.2 lbs CO_2/ ft^2-y
Error in credits Awarded	Because the project is awarded no credits the project will not be undertaken	Not Applicable	Unknown	Unknown

METHODOLOGY ASSESSMENT: This additional project is incorrectly identified as a free rider under the U.S proposal and is indeterminate under the EU proposal. Under the technology matrix, the project would not qualify, but may qualify under the project specific approach.

This project could possibly fall under two different main categories of the EU's positive list, energy efficiency or demand side management. The project may indeed qualify as additional under the second category. However, as we have seen, the first category, as currently defined, requires "significant improvements" to buildings. Thus far, the positive list offers no qualitative definition of the term "significant", making it impossible to determine whether this project would qualify under this category. The EU's positive list will need to assign a value to "significant" before it can be accepted as a viable approach to an international carbon offset program. Moreover, the fact that a single project could fall into two separate categories under the positive list and potentially result in different determinations of additionality is extremely problematic. It is imperative that categories and criteria used within each category included in the list be exclusive and clearly defined.

In other examples, we have seen how the U.S. approach will qualify a certain number of free rider projects; however, in this example, the opposite occurs – a truly additional project fails to qualify and because a primary goal of the project is to obtain credits, it will not be undertaken. Moreover, there are potential problems with the data for this project. To establish the U.S. proposal's threshold test, the data requirements turn on "recent and comparable" and at the moment these terms have been left undefined. To illustrate the problem, take the data from AAA Mill for this project. It can be argued that this facility is not "comparable" to the project facility. The two facilities are located in two different parts of the host country (latitude 37^0 South versus latitude 18^0 South) and possible face different climate conditions. In addition, there is a big difference in the energy used for heating by each facility (50,000 Btus/ft^2-yr versus 5,000 Btus/ft^2-yr). If the AAA Mill were excluded from the threshold test, it is possible that the project will qualify. Similar questions can be raised concerning the other benchmark facilities. For example, the steel mill is a corrugated iron building and may not be comparable to the concrete block pharmaceutical plant, even assuming both buildings are located in similar climate regimes. This example demonstrates that U.S. proposal needs to qualify the terms "recent and comparable" and that project qualification is totally dependent on the data points selected to establish the threshold i.e. the project may not qualify under one set of data points but may qualify under another set of data points.

Like the previous fly ash project, this insulation improvement project does not involve a new technology and is therefore rejected by the technology matrix. It demonstrates that the technology matrix approach will reject "low-tech" (e.g. replacement insulation, doors, or windows, weatherproofing, etc.) energy efficiency improvement projects for buildings even if the projects would result in true emissions reductions. However, the advantage of the technology matrix is that project developers are given the opportunity to fall back on the project-specific approach to qualify these types of projects. In this case, given the company's willingness to incur high transaction costs, it is likely that it would take advantage of this opportunity to qualify the project.

PROJECT NUMBER: IS9

COUNTRY: Jordan

SECTOR: Industrial

PROJECT TITLE: Highly Efficient Fertilizer Complex

PROJECT DESCRIPTION: This project involves the construction of a large, new fertilizer complex in the port city of Aqaba, Jordan. The project is made possible by the construction of a new natural gas pipeline connecting Aqaba with a Saudi gas field. The natural gas will be used by the fertilizer complex both as an energy source and as feedstock. The complex will have an annual capacity of 1 million tons of ammonium nitrate, 700,000 tons of nitric acid, 500,000 tons of urea, and 400,000 tons of ammonia.

The project is being designed to be as energy efficient as possible. New advanced technologies are being introduced into some of the plant processes. In particular, the ammonia production plant will utilize a new technology to reduce energy demand. (Ammonia is needed as an intermediate product in the production of nitrogenous fertilizer. The ammonia production process is endothermic, requiring much of the total energy needed to produce nitrogenous fertilizer.) However, for the most part the complex, including plants for the production of urea, ammonium nitrate, nitric acid, liquid nitrogen, ammonium sulfate, will utilize conventional processes and technologies, optimized to reduce energy demand throughout the production process.

The fertilizer complex will be a joint venture between a Jordanian company and a U.S. firm. Any credits generated by the project will be shared by the partners.

PROJECT ADDITIONALITY: The fertilizer complex is being built to take advantage of a new, inexpensive source of energy and feedstock in Aqaba, and to serve identified export markets via the Gulf of Aqaba, the Red Sea, and the Suez Canal. The push towards optimizing the energy efficiency of the complex is driven by a desire to reduce operating costs and hence secure a competitive advantage relative to older, existing fertilizer complexes serving the same markets. The decision to use an advanced, non-commercial technology in the ammonia production plant was driven by the same basic cost objective. The possibility of gaining credits was not factored into the design decisions, but the partners will nonetheless seek to qualify the complex for credits under an international carbon offset program. The project is therefore a free rider.

PROJECT EMISSIONS: The average energy demands of the fertilizer complex are expected to be 32 mmBtus per ton of product. The source of this energy will be a mix of natural gas, electricity, and distillate fuel; a weighted average emissions factor for these three sources has been computed at 180 lbs CO_2/mmBtu. Hence, the emissions rate (ER) of the complex can be computed as follows:

$$ER = (32 \text{ mmBtus/ton product})(180 \text{ lbs } CO_2/\text{mmBtu}) = 5,760 \text{ lbs } CO_2/\text{ton product}$$

PROJECT BENCHMARKS: We will assume that no new nitrogenous fertilizer complexes have been built in Jordan or the Middle East in the past 5 years, but that a total of four such complexes have been built worldwide. The available (fictional) data for these complexes are as follows:

Complex	Capacity, for Intermediate and Final Products (Million Tons/year)	Energy Used (Excluding Feedstock, mmBtus/ton)	Emissions Factor (lbs CO_2/mmBtu)	Emissions Rate (lbs CO_2/ton final product)
Uzbek A	Ammonium Nitrate—0.4 Nitric Acid—0.2 Urea—0.2 Ammonia—0.3*	60	170	10,200
ACME No. 1	Ammonium Nitrate—0.1 Nitric Acid—0.04 Urea—0.02**	10	190	1,900
AAA No. 2	Ammonium nitrate—0.7 Nitric Acid—0.5 Urea—0.4 Ammonia—0.2	37	180	6,660
ABC Plant	Ammonium nitrate—1.0 Nitric Acid—0.6 Urea—0.5 Ammonia—0.4	39	190	7,410
Average		146	182	6,542

*This complex produces more ammonia than is required for its own operation, and sells the excess to other Uzbek fertilizer production facilities.

**This complex does not have its own ammonia production plant, but buys ammonia from other producers.

114

PROJECT ANALYSIS TABLE: Project Number IS9

	U.S. Proposal	EU Proposal	Full Technology Matrix	Hybrid Technology Matrix
Does project qualify?	If "X" < 25th percentile then the threshold would be at least 10 mm Btus/ton product. The project's energy consumption of 32 mmBtus/ton product falls above the threshold and therefore would not qualify. Based on the available data, "X" would have to be set well above the 25th percentile in order for the project to qualify.	Positive list allows for energy efficiency projects using advanced technologies for and/or significant improvements in industrial processes. The project will use advanced technologies for ammonia production; thus, it would appear to qualify. However, the project is using an advanced non-commercial technology only for one section of the complex (ammonia production). This is a scenario that the positive list, as currently developed, is unable to manage. Thus, it cannot be determined whether or not the project would qualify.	The project is using an advanced non-commercial technology only for one section of the complex (ammonia production). This is a scenario that the technology matrix, as currently developed, is unable to manage. Therefore, it cannot be determined whether or not the project would qualify.	It cannot be determined whether or not the project would qualify.
Is the project correctly identified as either a free rider or an additional project?	Yes.	Indeterminate. Project creates a situation that the positive list is currently unequipped to handle. Project developers would be required to break the project into two separate projects.	Indeterminate. Project creates a situation that the technology matrix is currently unequipped to handle. Project developers would be required to break the project into two separate projects.	Indeterminate. Project creates a situation that the hybrid technology matrix is currently unequipped to handle. Project developers would be required to break the project into two separate projects.
Number of credits Awarded	Project does not qualify for credits	Not Applicable	Indeterminate	Indeterminate
Error in credits Awarded	Project is correctly identified as a free rider; thus, error in credits awarded is zero.	Not Applicable	Indeterminate	Indeterminate

METHODOLOGY ASSESSMENT: This free-rider project is correctly identified as such by the U.S. proposal, but is indeterminate under the EU proposal and the technology matrix approach.

Although the U.S. proposal correctly identifies the project as a free rider, there are potential problems with the data used to determine the threshold that need to be examined. First, the available data for this project only allows the threshold to be as low as the 25th percentile. However, this particular project, with only four recent and comparable facilities for comparison, demonstrates that for some projects the available data may not allow for a proper evaluation if "X" is a set percentile for all projects. For example, if "X" were set at the 10th percentile for all projects, a threshold could not be established for any project with less than 10 data points for comparison. This begs the question, should "X" be a set percentile for all projects or should "X" be based on the available data for individual projects. Second, the proposal's requirement for "recent and comparable facilities" presents other problems. In this case, no other fertilizer complexes had been built in the host country or within the host country's region, forcing a comparison to facilities in other parts of the world. This scenario likely is to be encountered throughout the developing world, as many developing nations will have undeveloped or underdeveloped industrial sectors. In addition, one of the fertilizer complexes used to develop the threshold test for this project is not engaged in ammonia production, which may exclude it as a "comparable" facility. Furthermore, none of the benchmark facilities produces intermediate and final products in the same proportion as the project facility. For large, complex industrial facilities producing numerous products, this is likely to be a very common situation. It is unclear that such facilities will have "comparable" let alone "recent" points of comparison, even if the search for such comparable facilities is carried out on a global basis.

This project creates a situation that both the technology matrix and positive list are unequipped, as currently developed, to handle. Up to this point, the positive list and technology matrix have been all or nothing approaches. In this case, however, only one of the plant processes (ammonia production) involves emission reductions from an advanced non-commercial technology. For the moment, both approaches are left with qualifying or not qualifying the entire fertilizer complex. Neither of these choices seems particularly appropriate in this situation. One solution would be to account for the emission reductions from the advanced non-commercial technology and qualify just that part of the project. In effect, under both the positive list and technology matrix, the project developers would be required to break this project down into two separate projects: the ammonia plant and the rest of the complex.

PROJECT NUMBER: IS10

COUNTRY: China

SECTOR: Industrial

PROJECT TITLE: Industrial Boiler Shutdown

PROJECT DESCRIPTION: A vehicle manufacturing facility in China has, for many years, operated its own coal-fired power plant to meet its own electricity needs, due to electricity reliability problems. However, due to the recent opening of a new foreign-owned hydropower plant in the area, as well as improvements in the grid's transmission capacity, these problems have been largely resolved. Furthermore, the price of coal has been rising faster than the price of electricity, due to increases in labor costs. Given these circumstances, the vehicle manufacturer has decided to shut down the on-site power plant, and to rely on the grid to meet its electricity needs.

The manufacturing facility is jointly owned by U.S. and Chinese firms, and the credits will be distributed among the partners based on their equity holdings.

PROJECT ADDITIONALITY: This project involves only a very limited capital expenditure, associated with the plant shutdown. It is being undertaken to reduce energy costs, not to gain credits. The project is a free rider.

PROJECT EMISSIONS: The facility's electricity demand is expected to remain unchanged by the plant shutdown; this demand has averaged 500 kWh per vehicle produced. The emissions factor for the facility will decline from 2.4 lbs CO_2/kWh to 0.9 lbs CO_2/kWh, because much of the power purchased from the grid is produced at the new hydropower plant. The project emissions rate (ER) can therefore be calculated as follows:

$$ER = (500 \text{ kWh/vehicle})(0.9 \text{ lbs } CO_2/\text{kWh}) = 450 \text{ lbs } CO_2/\text{vehicle}$$

PROJECT BENCHMARKS: Due to the chronic electricity supply problems characterizing China, 99 percent of existing manufacturing facilities have their own on-site electricity source, ranging from large steam turbine power plants to small diesel generators. During the past 5 years, no other existing facilities have shut down their on-site generators. Approximately 2 percent of new facilities opened in the last 5 years had no on-site generators.

The energy used by vehicle manufacturers in China averages 450 kWh per vehicle, and the emissions rate averages 900 lbs CO_2/vehicle.

PROJECT ANALYSIS TABLE: Project Number IS10

	U.S. Proposal	EU Proposal	Full Technology Matrix	Hybrid Technology Matrix
Does project qualify?	It is unclear whether the threshold test for the project should be based on a comparison of energy consumption or emission rates. Given that the energy consumption of the project is higher than the average consumption of all Chinese vehicle manufacturers, it is safe to assume that the project would not qualify even at "X" = 50th percentile. However, if the comparison is based on the emissions rate, the project's low rate relative to the sector average of 900 lbs CO2/vehicle would probably ensure that the project would qualify.	The project does not fit into any of the technology categories as they currently appear in the positive list; therefore, the project will not qualify under the EU proposal.	The project does not involve the use of an advanced, non-commercial technology and would not qualify under the technology matrix. The project developers would have the opportunity to qualify the project under the project specific approach. However, it appears to be a business as usual project and is unlikely to qualify under the project specific approach.	The project does not involve the use of an advanced, non-commercial technology and would not qualify under the technology matrix.
Is the project correctly identified as either a free rider or an additional project?	Indeterminate. Unclear whether to base threshold test on energy consumption or emission rates.	Yes.	Yes.	Yes.
Number of credits Awarded	Unknown	Not Applicable	The project does not qualify for credits.	The project does not qualify for credits.
Error in credits Awarded	Unknown	Not Applicable	The error in credits is zero.	The error in credits is zero.

METHODOLOGY ASSESSMENT: This free-rider project is correctly identified as such by the EU proposal and the technology matrix approach, but is indeterminate under the U.S. proposal

Under the U.S. proposal, industrial sector projects can establish the threshold through a comparison of either energy consumption or emissions rate. In this case, it is unclear which approach to use and the project's qualification status may differ depending on which one is used. If the threshold test is based on energy consumption, then the project would not qualify because the project will not effect the facility's electricity demand, which is already higher than the average of other facilities. It may qualify if the comparison is based on the emissions rate because the project's emissions would be significantly lower than the average emissions of other facilities. The possibility of different answers on the qualification issue in this particular case is clearly a problem that the U.S. proposal will need to address. This is a class of projects which does not fit into any of the categories around which the U.S. proposal was developed. In addition, the project's low emissions rate results not from the efforts of the plant operators, but from a new hydro plant built by someone else. Thus, if the project were to qualify, the project developers would be gaining credits for a hydro plant that they did not develop. Moreover, the choice of shutting down the coal plant was based on favorable economics rather than credit incentives.

PROJECT NUMBER: IS11

COUNTRY: South Africa

SECTOR: Industrial

PROJECT TITLE: Coal Mine Methane Recovery

PROJECT DESCRIPTION: Many coalbeds contain methane, which is released during the mining process. When present in the mine atmosphere in concentrations of 5 to 15 percent, methane is explosive. Therefore, mine operators are required by law to maintain methane concentrations well below 5 percent (e.g., 1 percent in the United States). Two main methods are utilized to reduce methane concentrations. First, fresh air is brought into the mine, via the mine's ventilation system, to dilute methane and keep its concentration below 5 percent. Secondly, vertical and horizontal wells are drilled into the coalbed in advance of, and subsequent to mining. When these wells are drilled in advance of mining, the gas retrieved from the wells is nearly 100 percent methane. This gas is often simply vented to the atmosphere. However, because of its relative purity, it can be recovered and sold as natural gas.

This project involves the recovery of methane from vertical wells drilled from the surface in advance of coal mining. The project is being financed jointly by the South African coal company and a U.S. company specializing in coalbed methane recovery. The latter company will receive any credits awarded to the project, along with a share of the recovered methane, in exchange for its financial contribution.

PROJECT ADDITIONALITY: Coalbed methane recovery is currently being utilized in the United States, and we assume, for the purpose of this case study, that it is also being utilized at various mines in South Africa. However, in the past, this particular mine simply vented the methane, because the cost of recovery exceeded the value of the methane. However, with the potential for gaining credits, the mining company has reassessed the project's economic feasibility. It has been determined that, with the credits, the project is economically viable. Therefore, the project is not a free rider.

PROJECT EMISSIONS: This large mining operation is considered very gassy, with a high methane emissions rate. Prior to the project, the surface wells emitted methane at an average rate of 100 cubic feet per minute (cfm). Subsequent to the project, all of this gas will be recovered. However, it is estimated that approximately 2 percent of the recovered gas will eventually leak at various points between the wells and the final end users. Therefore, the project's emissions rate (ER) is estimated as follows:

ER = (0.02)(100 cfm)(0.0422 lbs methane/cubic foot) (1 ton/2000 lbs) (525,600 minutes/yr)

ER = 22 tons methane/year

PROJECT BENCHMARKS: Let us assume that there are 10 other coal mines in South Africa that drill wells in advance of mining. Relevant (fictional) data for these mines are as follows:

Mine	Handling of Methane from Surface Wells	Average Emissions Rate (Tons Methane/Year)
No. 1	Vented	100
No. 2	Vented	250
No. 3	Vented	40
No. 4	Recovered	15
No. 5	Recovered	10
No. 6	Vented	500
No. 7	Vented	1,000
No. 8	Vented	700
No. 9	Recovered	25
No. 10	Vented	1,500
Average		414

PROJECT ANALYSIS TABLE: Project Number IS11

	U.S. Proposal	EU Proposal	Full Technology Matrix	Hybrid Technology Matrix
Does project qualify?	The U.S. approach states, "if methane capture was not a standard practice, then the threshold would be any capture project that is better than the current situation." If methane capture is a standard practice, then it is assumed that the project activity is required to meet the "significantly better than average" threshold. However, the proposal does not define what is meant by "standard" practice and it is therefore not possible to establish a threshold for additionality determination.	This project qualifies as additional under the EU's positive list. However, it falls under 2 categories on the list: 1) the sub-category of "significant improvements in industrial processes . . ." of the energy efficiency category, and 2) "significant improvements in existing energy production.	Coalbed methane recovery is being used at various mines in South Africa. Thus, this is not an advanced, non-commercial technology, and as such, it will not qualify as additional under the technology matrix. Project developers would have the opportunity to qualify the project under the project-specific approach. Given that in the past, this mine did not recover methane due to the high cost, this project would likely qualify.	Coalbed methane recovery is being used at various mines in South Africa. Thus, this is not an advanced, non-commercial technology, and as such, it will not qualify as additional under the technology matrix. Project developers would have the opportunity to qualify the project under the project-specific approach, and given that in the past, this mine did not recover methane due to the high cost, this project would likely qualify.
Is the project correctly identified as either a free rider or an additional project?	Indeterminate. Additionality could not be determined due to the lack of a definition of "standard practice"	Yes.	Yes, if project developers choose to utilize the project specific approach.	Yes, if project developers choose to utilize the project specific approach.
Number of Credits Awarded	Unknown	Not applicable.	As discussed in a previous report,* the estimation of methane reductions from mining projects do not require benchmarks. Instead, the metered amount of methane recovered from pre-mining and post-mining wells will provide an estimate of methane reductions.	No benchmark is needed. Credits would be calculated based on the amount of gas captured. In this case, credits will be awarded at a rate of 22 tons methane/year.
Error in Credits Awarded	Unknown.	Not applicable.	Unknown.	Unknown.

*SAIC, *"Developing the Technology Matrix for India and Ukraine,"* Draft Report, August 2000, pg. 80

122

METHODOLOGY ASSESSMENT: This additional project was correctly identified as such by all approaches except for the U.S. approach. Under the U.S. approach, methane capture projects are required to meet the "significantly better than average" threshold. If methane capture is not standard practice, then "the threshold would be any capture project that is better than the current situation." However, the U.S. approach does not explain what is meant by "standard practice" and "better than the current situation". As a result, the recovery project is indeterminate when applied to the U.S. approach. In this project scenario, coalbed methane recovery is being utilized at various mines in South Africa. Three out of the 10 listed reference projects comprise recovery projects. In the remaining seven mines, the methane is vented. If the cut-off for standard practice is half of the projects examined, the project would qualify as additional. However, if the cut-off is 25 percent, the recovery project would not be additional. This issue of how to determine standard practice would be even more difficult in those cases where it is impossible to obtain adequate data on mining practices. In any case, a more precise definition of the evaluation criteria must be outlined, before the U.S. approach can be applied effectively to methane capture projects.

The U.S. approach indicates that the baseline for methane projects would "be the previously existing condition -- i.e., no capture -- and the project would calculate credits based on the amount of gas captured." Hence, no benchmark is needed for estimating the credits awarded to the project. Simply, the credits awarded would be based on the amount of captured gas metered at the mine. In this case, 22 tons of methane each year.

This project qualifies as additional under the EU's positive list, as it entails a "significant improvement in an industrial process." The project could also qualify under the category of "significant improvements in existing energy production". While the term "significant" should be clarified and further defined; it is assumed that the reduction in emissions rate due to the project is considered "significant" (emissions rate of the project is 50 times less than the current emissions rate). However, more importantly, the positive list should be specified in more detail to avoid the problem of overlapping project categories.

While the project does not initially qualify under the full and hybrid technology matrix approaches because the project does not involve an advanced, non-commercial technology, the project developers may use the project specific approach to determine additionality. In this case, it is likely that the project would qualify using the project specific approach, particularly, because in the past, the mine did not recover methane due to a very high cost.

In a previous report, *"Developing the Technology Matrix for India and Ukraine, Draft Report,"* (August 2000), it was concluded that the estimation "of methane reductions resulting from a mining-associated project is unusual, and does not require a benchmark." Instead of estimating the methane reductions from such a project, it is recommended that the metered amount of methane recovered from the pre-mining and post-mining wells be used to provide an estimate of methane reductions. Thus, for a coal bed methane recovery project such as this, there is no need to develop a benchmark with which to estimate methane reductions. Instead, such reductions can easily be metered.

PROJECT NUMBER: IS12

COUNTRY: Argentina

SECTOR: Industrial

PROJECT TITLE: Landfill Gas Flaring

PROJECT DESCRIPTION: In the United States, regulations were recently implemented requiring landfills with more than 2.5 million metric tons of waste in place, and annual emissions of nonmethane volatile organic compounds (NMVOCs) greater than 50 metric tons, to collect and burn their landfill gas emissions. In this project, we imagine a future in which similar regulations have been implemented in Argentina.

The project involves the collection and flaring of gas from a large landfill subject to the new regulatory requirements. Prior to the project, the gas from the landfill was simply vented. Most of the project financing will be provided by the landfill operator. However, some additional funding will be provided by a U.S. company, in exchange for the credits to be generated by the project.

PROJECT ADDITIONALITY: Although the collection and flaring of landfill gas is a fairly common practice in the United States and other areas, this project represents the first time the technology will be applied to an Argentine landfill. The project is being undertaken not to obtain credits, but because the landfill operator is required to do so under the new regulations. Therefore, the project is a free rider.

PROJECT EMISSIONS: Prior to the project, the landfill emitted an average of 25 million cubic feet of methane per year. The flaring project will convert these methane emissions into carbon dioxide emissions. Using the standard methane emissions factor of 115.3 lbs CO_2 per mmBtu, the project's CO_2 emissions rate (ER) can be estimated as follows:

ER = (25 million cubic feet/yr)(1000 Btus/cubic foot)(115.3 lbs CO_2/mmBtu) (1 ton/2000 lbs)

ER = 1440 tons CO_2 per year

PROJECT BENCHMARKS: We will assume that there are approximately 3000 other landfills located throughout Argentina, all of which produce some methane and none of which currently utilize landfill gas recovery technologies. The average emissions from these landfills are estimated at 6 million cubic feet of methane per year.

PROJECT ANALYSIS TABLE: Project Number IS12

	U.S. Proposal	EU Proposal	Full Technology Matrix	Hybrid Technology Matrix
Does project qualify?	The U.S. approach states, "if methane capture was not a standard practice, then the threshold would be any capture project that is better than the current situation." In this case, methane capture at landfills is not a standard practice in Argentina. However, it is unclear what data to compare this project to in order to determine additionality. Thus, additionality is indeterminate.	The positive list allows "advanced technologies or significant improvements in industrial processes" to qualify. It is unclear whether gas flaring and methane recovery at landfills constitute industrial processes. It is also unclear whether flaring of landfill gas is an advanced technology, as it is common in the U.S. and other areas.	This project represents the first time this technology will be applied to an Argentine landfill; thus, this project qualifies as additional under the technology matrix, as it represents an advanced, non-commercial technology.	This project represents the first time this technology will be applied to an Argentine landfill; thus, this project qualifies as additional under the technology matrix, as it represents an advanced, non-commercial technology.
Is the project correctly identified as either a free rider or an additional project?	Indeterminate. Although it is clear that methane capture is not a standard practice in Argentina, the U.S. methodology does not offer clear guidance on what data to compare the project to determine additionality (i.e. the definition of "any capture project" is insufficient).	Indeterminate. Positive list concepts and terms must be more clearly defined.	No. Market penetration and economic feasibility tests do not account for the fact that the project is required, under new regulations, and will be conducted with or without the carbon offset program.	No. Market penetration and economic feasibility tests do not account for the fact that the project is required under new regulations, and will be conducted with or without the carbon offset program.
Number of Credits Awarded	No benchmark is required. Credits would be calculated based on the amount of gas flared. In this case, credits will be awarded at a rate of 1440 tons of CO_2 per year.	Not applicable.	As discussed in a previous report,* the estimation of methane reductions from gas flaring and recovery projects do not require benchmarks. Instead, the metered amount of methane recovered from the landfill wells will provide the estimate of methane reductions.	No benchmark is required. Credits would be calculated based on the amount of gas flared. In this case, credits will be awarded at a rate of 1440 tons of CO_2 per year.
Error in Credits Awarded	Unknown.	Not applicable.	Project is a free rider, so error in credits awarded = 100%.	Project is a free rider, so error in credits awarded = 100%.

*SAIC, "*Developing the Technology Matrix for India and Ukraine,*" Draft Report, August 2000, pg. 80.

METHODOLOGY ASSESSMENT: This free rider project failed to be correctly identified as such under all four approaches. None of the tests, as currently defined, addresses the possibility that a project might be required under existing regulations. All four tests should include a criteria that projects required under law be automatically disqualified for credits, as such projects will invariably be free riders.

There are other problems with the additionality tests as well. As previously discussed, the U.S. approach states that "if methane capture at landfills was not standard practice, then by definition the threshold would be any capture project that is better than the current situation." However, the U.S. methodology does not offer guidance on what data to compare the project to in order to establish the threshold. Thus, an evaluation of the project under the current U.S. approach finds the additionality of the project indeterminate.

An evaluation of the project under the EU's positive list finds the additionality of the project indeterminate, due to the lack of clarity and clearly defined positive list concepts and terms. The positive list allows advanced technologies or significant improvements in industrial processes to qualify. However, it is unclear whether gas flaring and recovery at landfills would constitute an "industrial" process. Further, it is unclear whether flaring of landfill gas would constitute an advanced technology, as it is a common practice in the U.S. and other areas. Finally, it is unclear whether this project constitutes a "significant" improvement in an industrial process. Such terms and concepts under the EU's positive list must be clarified in order to avoid indeterminate additionality classifications.

This free rider project was incorrectly identified as additional under both technology matrix approaches. Because the project represents the first time this technology will be applied to an Argentine landfill, it will qualify as additional under the technology matrix. Such first-of-its-kind projects do not necessarily always come into being due to an economic subsidy such as credits. In this particular case, the landfill operator is required to undertake the project due to new regulations. However, because it is applied on a country-by-country basis, the market penetration test will of necessity always qualify first-of-its-kind projects as additional.

As discussed in the previous case study (IS11), it was concluded in a previous report on *"Developing the Technology Matrix for India and Ukraine,"* that the estimation of methane reductions for landfill methane flaring and recovery projects does not require a benchmark. Instead of estimating the methane reductions from such a project, it is recommended that the metered amount of methane recovered from the landfill wells be used to provide an estimate of methane reductions. In essence, the estimation of recovered methane is an issue of project metering and monitoring; it is not a benchmark issue. Thus, for a landfill gas flaring and recovery project such as this, there is no need to develop a benchmark with which to estimate methane reductions -- such reductions can easily be metered at the project site.

PROJECT NUMBER: IS13

COUNTRY: Kazakhstan

SECTOR: Industrial

PROJECT TITLE: Recovery of Associated Natural Gas

PROJECT DESCRIPTION: This project involves the construction of a 200-mile natural gas transmission line to serve an existing oil field. Currently, the gas associated with the oil is being flared. Once the pipeline has been completed, the natural gas will be recovered and sold in both domestic and foreign markets. A consortium of U.S. and Kazak oil companies is developing the oil field. The consortium will finance the pipeline, and distribute any credits awarded to the project to the consortium members according to their equity shares in the oil field.

PROJECT ADDITIONALITY: This is a major project that is being undertaken to recover and sell a valuable product that was previously being wasted. The project is viable without credits, and the value of any credits that might be awarded to the project was not taken into account when the economic feasibility assessment of the project was performed. The project is therefore a free rider.

PROJECT EMISSIONS: The oil field produces an average of 200 million cubic feet of associated gas per year. Previously, all of this gas was flared, yielding carbon dioxide emissions of 12,070 tons per year. However, the project now captures all of the associated gas, with an estimated leakage of only 2 percent. Therefore, the project's emission rate can be estimated as follows:

$$ER = (0.02)(200,000 \text{ thousand cubic feet/yr})$$
$$(42.2 \text{ lbs methane/thousand cubic foot})(1 \text{ ton/2000 lbs})$$

$$ER = 84 \text{ tons methane}$$

PROJECT BENCHMARKS: Oil field emissions data are unavailable for Kazakhstan and other FSU countries.

PROJECT ANALYSIS TABLE: Project Number IS13

	U.S. Proposal	EU Proposal	Full Technology Matrix	Hybrid Technology Matrix
Does project qualify?	The threshold for methane capture requires activities to be significantly better than the average or better than the current situation if the project activity is not standard practice. However, for this project it is unclear if recovery of associated gas is standard practice in Kazakhstan and it cannot be determined whether or not the project would qualify.	The positive list allows for significant improvements in existing energy production. However, the definition of "significant" is unclear; therefore, project qualification cannot be determined.	This project involves the use of a commercial technology that is available worldwide. Therefore, it does not qualify. Under the technology matrix, project developers may use the project-specific approach to attempt to qualify the project, but since it is a free rider, it is unlikely that additionality could be demonstrated.	This project involves the use of a commercial technology that is available worldwide. Therefore, it does not qualify. Under the technology matrix, project developers may use the project-specific approach to attempt to qualify the project, but since it is a free rider, it is unlikely that additionality could be demonstrated.
Is the project correctly identified as either a free rider or an additional project?	Indeterminate. Information on Kazakhstan's oil production processes is unavailable.	Indeterminate	Yes	Yes
Number of Credits Awarded	Indeterminate	Not Applicable	None	None
Error in Credits Awarded	Unknown	Not Applicable	Zero. The project is correctly identified as a free rider.	Zero. The project is correctly identified as a free rider.

128

METHODOLOGY ASSESSMENT: This free rider project is indeterminate under the U.S. and EU proposals but is correctly identified as such under the technology matrix.

The U.S. proposal as it relates to methane capture is incomplete. It stipulates that methane capture projects are required to meet the "significantly better than average" threshold but also indicates that if the methane capture process is not standard practice in the host country then the threshold is any capture project that is better than the current situation. In other words, it appears that if the project's methane capture process is not normally utilized in the host country then any project using that process could potentially be used to establish the threshold. If the methane capture process is in use, we then assume that a percentile distribution of recent and comparable activities would need to be developed. Unfortunately, the section on methane capture is not well defined, and we are assuming that the proposal takes a kind of dual threshold test approach to methane capture projects. The U.S. proposal will need to further define what it means by "standard practice." Would standard practice be based on a certain market penetration percentage within the host country? Furthermore, even if the natural gas recovery system to be used in the project were deemed standard practice, the information needed to establish a threshold and perform the percentile test is not only unavailable for Kazakhstan but for the rest of the FSU as well. Once again, this lack of data reveals one of the potential problems with the U.S. proposal, i.e. its data requirements are often more advanced than what is available on a country or even regional level.

With respect to defining terms, a similar problem exists for the EU proposal. Under its energy efficiency category, the positive list qualifies "advanced technologies for and/or significant improvements in industrial processes" and "significant improvements in existing energy production." In the absence of qualifiers for the term "significant," we cannot determine if this natural gas recovery project qualifies as a significant improvement to an industrial process and/or a significant improvement in existing energy production. In addition, this particular project highlights confusion over the term "advanced technologies." Natural gas recovery systems are in use worldwide, but their use in Kazakhstan currently is unknown. If these recovery systems are a new technology for a country like Kazakhstan, is this a sufficient criterion to meet the positive list's advance technology criteria? The EU should specify whether positive list technologies/processes are to be global or country-specific. Further refinement of the EU proposal is clearly needed for it to be a legitimate screen for project qualification.

As for the technology matrix, the project involves a commercial technology and therefore would not qualify. The project developers would be afforded the opportunity to qualify the project under the project-specific approach; however, qualification under this approach is unlikely because the project is viable without the incentive of emission reduction credits.

PROJECT NUMBER: TS1

COUNTRY: India

SECTOR: Transportation

PROJECT TITLE: Dedicated Compressed Natural Gas (CNG) Taxis

PROJECT DESCRIPTION: The city of New Delhi, India is one of the world's ten most polluted cities. Almost 70 percent of the air pollution in the area can be attributed to vehicular emissions. In response, the government of India has introduced several initiatives to promote cleaner transportation. The Supreme Court directed the city to increase the number of CNG refueling facilities in Delhi from 9 to 80 and to convert all diesel buses to CNG by March 31, 2000. The natural gas used for the CNG vehicles will be supplied by a pipeline, which has already been built from the West coast of India to New Delhi.

Encouraged by city plans to build a refueling station near the local fleet station, a local taxi fleet operator has decided to replace his entire fleet of 150 gasoline fueled vehicles (average age of 10 years) with new dedicated CNG vehicles. Moreover, the operator is planning to offset some of the high cost of purchasing natural gas vehicles by selling the future GHG emission reduction credits associated with switching from conventional gasoline to CNG vehicles. The CNG vehicles will be imported from a U.S.-based original equipment manufacturer (OEM), who will provide a discount and low-interest loans in exchange for a majority of the credits awarded for the project.

PROJECT ADDITIONALITY: At least 12,000 conventional vehicles have been converted for CNG use in Delhi, Mumbai, and Baroda. However, there are no dedicated CNG vehicles in India. Dedicated CNG vehicles built by OEMs are considerably more expensive than conventional gasoline vehicles. Although some of the high up-front costs of purchasing CNG vehicles can be offset by the lower fuel cost of natural gas, dedicated CNG passenger cars are still an advanced technology that requires subsidies or other assistance to be implemented in India. Moreover, the availability of carbon offset program credits helped convince the taxi fleet operator to go ahead with the project. Therefore, the project is additional.

PROJECT EMISSIONS: The CO_2 exhaust emission factor of the CNG vehicles is 136.24 g CO_2/km.[14] The CH_4 emission factor is 0.7 g CH_4/km. The global warming potential (GWP) of CH_4 is 21.[15] The project's emission rate is as follows:

ER for CO_2 = 136.24 g CO_2/km

---

[14] These emission factors only include exhaust emissions. GHG emissions during the production, transportation, and distribution of the fuels are not included in this equation.
[15] Intergovernmental Panel on Climate Change (IPCC), *Revised 1996 IPCC Guidelines for National Greenhouse Gas Inventories,*" 1997.

$$\text{ER for } CH_4 = (0.7 \text{ g } CH_4/\text{km})(21) = 18.9 \text{ g } CO_2 \text{ equivalent/km}$$

$$\text{ER project} = 136.24 \text{ g } CO_2/\text{km} + 8.9 \text{ g } CO_2 \text{ equivalent/km} = 155 \text{ g } CO_2 \text{ equivalent/km}$$

PROJECT BENCHMARKS: No data is available describing exhaust emissions of carbon dioxide and methane from different types of road transportation vehicles in India. Instead, we have used fictional average emission factors for the different models of new passenger cars sold in India in 1998. The variation in emission factors is included to facilitate benchmark development and has no relation to the actual performance of individual vehicle models. As no dedicated natural gas vehicles are sold in New Delhi, all of the vehicle models included are powered by conventional fuel (gasoline and diesel).

Estimated Emission Factors for New Passenger Vehicles Sold in India, 1998						
Control Technology	Fuel Type[16]	Number of Vehicles Sold[17]	Share of Total Vehicles Sold (%)	Emissions (g/km)		
				CO_2	CH_4	Total CO2 Equivalent
Model 1	GL	9,208	2.29	285	0.025	285.5
Model 2	GL	7,408	1.84	282	0.027	282.6
Model 3	GL	483	0.12	286	0.029	286.6
Model 4	GL	8,258	2.05	286	0.029	286.6
Model 5	GL	8,448	2.10	291	0.037	291.8
Model 6	GL	3,542	0.88	295	0.038	295.8
Model 7	GL	273,672	68.05	296	0.038	296.8
Model 8	GL	3,573	0.89	293	0.037	293.8
Model 9	GL	452	0.11	292	0.037	292.8
Model 10	GL	392	0.10	298	0.040	298.8
Model 11	GL	3,437	0.86	303	0.041	303.9
Model 12	GL	1,349	0.34	284	0.028	284.6
Gasoline Total		**320,222**	**79.63**			
GL Weighted Average						**295.7**
Model 13	DL	15,283	3.80	298	0.040	298.8
Model 14	DL	3,875	0.96	301	0.041	301.9
Model 15	DL	59,964	14.91	305	0.042	305.9
Model 16	DL	239	0.06	288	0.031	288.7
Model 17	DL	2,560	0.64	310	0.043	310.9
Diesel Total		**81,921**	**20.37**			
Diesel Weighted Average						
Average						294.5
Weighted Average				303	0.039	297.5
Total		402,143	100			

---

[16] Distinction between gasoline and diesel powered vehicles is fictional and included only for the purpose of baseline development

[17] 1999 Ward's Automotive Yearbook

PROJECT ANALYSIS TABLE: Project Number TS1

	U.S. Proposal	EU Proposal	Full Technology Matrix	Hybrid Technology Matrix
Does project qualify?	The project will qualify even if "X"< 1st percentile, or lower than 282.6 (ER<282.6).	Project qualifies as additional under the EU's positive list, because it involves more efficient and less polluting modes of transportation, and improves or substitutes existing vehicles.	The project deploys an advanced technology that is not yet fully commercial and does not have a market penetration in India. The project will qualify as additional.	The project qualifies as additional under the technology matrix. Thus, it will also qualify under the hybrid approach.
Is the project correctly identified as either a free rider or an additional project?	Yes	Yes	Yes	Yes
Number of credits Awarded	The U.S. approach does not clarify whether credits are awarded by comparing the project to a sector/fuel average, or the emission factor of the activity to be replaced. In this case, the project credits are determined by subtracting the project ER from the ER of the gas vehicles to be replaced* (credits=508.7-155.14=353.56 g CO_2 eq/km).	Not applicable	In an earlier study, we recommend that credits be derived for replacement transportation projects by subtracting the ER of the vehicles to be replaced (155.14) from the ER of the project. This study did not address what information can be used to represent emissions of the vehicles to be replaced. In this case, the ER of the old vehicles is unknown. To estimate the old vehicle ER, we use the IPCC default emission rate for vehicles in developing countries (506 g CO_2/km and 0.13 g CH_4/km).* Thus, credits awarded=508.7-155.14=353.56 g CO_2 eq/km	The project credits would be determined by subtracting the project ER from the ER of the gasoline vehicles to be replaced (credits=508.7-155.14=353.56 g CO_2 eq/km
Error in credits Awarded	Unknown	Not applicable	Unknown	Unknown

* According to the IPCC, older passenger vehicles in the developing world are typically equipped with the lowest IPCC category of control technology. The average emission rate for "uncontrolled" passenger vehicles is 506 g CO_2/km and 0.13 g CH_4/km. This would equal a total emission rate of 508.7 g CO_2 equivalent/km. Intergovernmental Panel on Climate Change (IPCC), "*Revised 1996 IPCC Guidelines for National Greenhouse Gas Inventories*," 1997.

METHODOLOGY ASSESSMENT: This project is the first example of a transportation project. The U.S. approach does not provide guidance on benchmark development for the transportation sector. However, according to the U.S. approach for other sectors, the additionality test should involve a comparison with a reference scenario consisting of a set of recent and comparable activities. In this case, we interpreted "recent and comparable" as meaning new passenger vehicles sold in 1998. By using this data and applying the U.S. approach, the CNG project would qualify as additional regardless of the percentile threshold selected.

This project involves a replacement project based on a fuel switch from gasoline to natural gas. The US guidance on estimating credits from replacement versus new capacity projects is unclear and should be clarified. According to the U.S. methodology, it appears that credits would be awarded for a replacement project in the power sector by subtracting the project emissions from the emissions of the activity to be replaced. Projects that include new activities or provide new generation would use a sector average (or fossil average in the case of the power sector). This project is a replacement project. Thus, we subtracted the emissions rate of the project from the emissions rate of the vehicles to be replaced. In this case, the IPCC default emission factor for older vehicles in developing countries was used to estimate the potential credits. Accordingly, the U.S. approach would award credits at a rate of 353.56 g CO_2 equivalent/km. However, the U.S. methodology should be clarified to provide more detailed guidance on how to estimate credits from replacement projects versus new capacity projects.

As transportation projects that improve efficiency or substitute existing vehicles are included on the EU positive list, this project also qualifies as additional under the EU's approach. The two technology matrix approaches would also qualify the project. In this case, the technology matrix awards credits at a rate of 353.56 g CO_2 equivalent/km. However, the guidelines for estimating emission reduction credits under the technology matrix should be more detailed in terms of specifying which data can be used to represent emissions of the vehicles to be replaced. In this case, a global default emission factor was used.

Nonetheless, CNG replacement vehicle projects appear to represent a class of projects are handled equally well by all methods. The four baseline approaches correctly identify the project as additional and the credits awarded are the same for the U.S. and technology matrix approaches.

PROJECT NUMBER: TS2

COUNTRY: India

SECTOR: Transportation

PROJECT TITLE: New Gasoline-Fueled Taxis

PROJECT DESCRIPTION: Many urban areas throughout Eastern and Central Europe, Asia, Africa and Latin America are experiencing massive population growth due to economic expansion and migration from rural areas. This growth has led to an equal rise in transportation needs and private vehicle ownership. In New Delhi, India the transportation sector is growing by 7 percent annually. Much of this growth stems from a rise in private vehicle ownership. The city's public transportation system continues to deteriorate, even though the demand for public transportation is growing rapidly.

In response to the growing transportation needs, a local taxi fleet operator is planning to expand his taxi fleet by 100 vehicles. The fleet operator is considering three options: used vehicles, new compressed natural gas (CNG) vehicles, and new conventional gasoline vehicles. Due to the high transportation demand, the operator is hesitant about purchasing used vehicles that are likely to require increased maintenance and will go out of service faster. Purchasing new CNG vehicles has also been ruled out. There is no CNG refueling stations near the service area of the taxi fleet. Hence, the operator would have to install the necessary refueling infrastructure in addition to investing in new vehicles, making the fleet expansion too costly.

Instead, the fleet operator has decided to purchase new gasoline fueled vehicles of a European manufacturer (Model 12). Although more costly than used vehicles, Model 12 would be more reliable in the long run, the fuel economy is excellent, and the vehicle would be easy to refill at any of New Delhi's many gasoline stations. In addition, the European parent company of the enterprise producing and marketing Model 12 in India, is interested in investing in the project in exchange for the potential carbon offset program credits. The European car maker hopes that by participating in the project, the company might advance their product, boost sluggish sales, and gain market advantage in the traditionally Indian dominated automobile market.

PROJECT ADDITIONALITY: The project is a free-rider. The primary objective of the project developer (i.e., the fleet operator) is to find the most cost-effective and reliable method for expanding capacity in response to increased demand. The investors are mainly interested in the project as a measure of promoting their product in the Indian market. Hence, the project would occur even without the prospect of carbon offset program participation.

PROJECT EMISSIONS: The emissions rate of the 2001 Model 12 is somewhat better than previous model years. The CO_2 exhaust emission factor of the gasoline vehicles is

180 g CO_2/km.[18] The CH_4 emission factor is 0.026 g CH_4/km. The global warming potential (GWP) of CH_4 is 21.[19] The project's emission rate is as follows:

ER for CO_2 = 283 g CO_2/km

ER for CH_4 = (0.028 g CH_4/km)(21) = 0.6 g CO_2 equivalent/km

ER project = 283 g CO_2/km + 0.6 g CO_2 equivalent/km = 283.6 g CO_2 equivalent/km

PROJECT BENCHMARKS:

No data is available describing exhaust emissions of carbon dioxide and methane from different types of road transportation vehicles in India. As in the first transportation project, we have used fictional average emission factors for the different models of new passenger cars sold in India in 1998. The variation in emission factors is included to facilitate benchmark development and has no relation to the actual performance of individual vehicle models. All of the vehicle models included are powered by conventional fuel (gasoline and diesel).

---

[18] These fictional emission factors only include exhaust emissions. GHG emissions during the production, transportation, and distribution of the fuels are not included in this equation.

[19] Intergovernmental Panel on Climate Change (IPCC), *Revised 1996 IPCC Guidelines for National Greenhouse Gas Inventories,*" 1997.

Estimated Emission Factors for New Passenger Vehicles Sold in India, 1998						
Control Technology	**Fuel Type**[20]	**Number of Vehicles Sold**[21]	**Share of Total Vehicles Sold (%)**	Emissions (g/km)		
				CO_2	**CH_4**	**Total CO2 Equivalent**
Model 1	GL	9,208	2.29	285	0.025	285.5
Model 2	GL	7,408	1.84	282	0.027	282.6
Model 3	GL	483	0.12	286	0.029	286.6
Model 4	GL	8,258	2.05	286	0.029	286.6
Model 5	GL	8,448	2.10	291	0.037	291.8
Model 6	GL	3,542	0.88	295	0.038	295.8
Model 7	GL	273,672	68.05	296	0.038	296.8
Model 8	GL	3,573	0.89	293	0.037	293.8
Model 9	GL	452	0.11	292	0.037	292.8
Model 10	GL	392	0.10	298	0.040	298.8
Model 11	GL	3,437	0.86	303	0.041	303.9
Model 12	GL	1,349	0.34	284	0.028	284.6
Gasoline Total		**320,222**	**79.63**			
GL Weighted Average						**295.7**
Model 13	DL	15,283	3.80	298	0.040	298.8
Model 14	DL	3,875	0.96	301	0.041	301.9
Model 15	DL	59,964	14.91	305	0.042	305.9
Model 16	DL	239	0.06	288	0.031	288.7
Model 17	DL	2,560	0.64	310	0.043	310.9
Diesel Total		**81,921**	**20.37**			
Diesel Weighted Average						
Average						**294.5**
Weighted Average				**303**	**0.039**	**297.5**
Total		**402,143**	**100**			

---

[20] Distinction between gasoline and diesel powered vehicles is fictional and included only for the purpose of baseline development

[21] 1999 Ward's Automotive Yearbook

PROJECT ANALYSIS TABLE: Project Number TS2

	U.S. Proposal	EU Proposal	Full Technology Matrix	Hybrid Technology Matrix
Does project qualify?	The project will qualify if "X"< 2nd percentile, or lower than 284.6 (ER<284.6).	According to the EU positive list, "projects will qualify if they involve more efficient and less polluting modes of transportation, and improve or substitute existing vehicles." The text does not clarify what is meant with "improve or substitute". As a result, it can not be determined whether the project will qualify.	The project does not apply an advanced technology. Similar models of the proposed vehicles are already being sold on the Indian market although at a low percentage (0.34). The project will not qualify as additional under the technology matrix. The project developers would be given the opportunity to prove additionality using the project-specific approach. However, given that the project represents a regular capacity expansion effort, it is unlikely that additionality could be proved.	The project does not qualify as additional under the technology matrix. Thus, it will not qualify under the hybrid approach.
Is the project correctly identified as either a free rider or an additional project?	No. U.S. approach does not provide guidance on benchmark development for the transportation sector. If the percentile threshold test developed for the electricity and industrial sectors is applied, the project qualifies.	Indeterminate. Vehicle model in this example does have a better emission rate compared to previous models, but the positive list does not clarify what is meant by "improvement or substitution of existing vehicles."	Yes	Yes
Number of credits Awarded	As this is a new capacity project, the credits would be determined by subtracting the project ER from the average ER of the recent, comparable vehicles, in this case gasoline vehicles (credits=295.7-283.6=12.1 g CO_2 eq/km)	Not applicable	Project does not qualify for any credits	Project does not qualify for any credits
Error in credits Awarded	The project is a free rider; therefore the error is equal to 100 percent of the credits awarded (12.1 g CO_2 eq/km)	Not applicable	The project is correctly identified as a free rider; therefore the error in the credits awarded is zero.	The project is correctly identified as a free rider; therefore the error in the credits awarded is zero.

METHODOLOGY ASSESSMENT: This project only qualifies for credits under the U.S. approach. As mentioned in the previous transportation case study, the U.S. approach does not provide guidance on benchmark development for the transportation sector. If we apply the "recent and comparable" additionality test similarly to the method used in the previous case study, the project will qualify for credit even if the threshold is set at the 2nd percentile (ER<284.6) and will be awarded credit at a rate of 12.1 g CO_2 eq/km. This result was achieved by comparing the project emissions rate to the average emissions rate of new passenger vehicles sold in India in 1998. Thus, a vehicle model, which is already available on the Indian automobile market, will be able to receive credits under the U.S. baseline approach. As CDM and JI projects, by definition, must include foreign participation, the qualification of projects like this new gasoline vehicle project raises the question whether the U.S. methodology may inadvertently favor or subsidize foreign conventional technology and investment projects in India relative to projects utilizing domestic technologies and investment.

It is not clear whether this type of transportation project would receive credit under the EU positive list. In the EU's proposed positive list of technologies it is stated that projects that lead to the "improvement or substitution of existing vehicles" will qualify for credit. However, the proposal does not clarify what is meant by "improvement" of existing vehicles. The vehicle model included in this project does have a better emission rate compared to that of previous model years. Yet, it is unclear whether this improvement is enough to allow the project to qualify for credits under the EU positive list.

The two technology matrix approaches would disqualify this project because it applies conventional vehicle technology that has already been introduced in India. As such, the matrix approaches are better than the U.S. approach at screening out conventional technology free riders.

COUNTRY: China

SECTOR: Transportation

PROJECT TITLE: Aluminum Rail Cars for Efficient Coal Transport

PROJECT DESCRIPTION: Most of China's coal reserves are located in the North-Western provinces, while many of the power plants using the coal are located in the more populated provinces in the South-East. Thus, enormous amounts of coal are transported daily across the country's rail system. This project involves the purchase of aluminum rail cars, instead of steel cars, to reduce the weight and diesel fuel use of freight trains. It is estimated that the use of aluminum to build freight cars can reduce the weight of each car by 30-40 percent. The project is being undertaken by a large state-owned utility in the South. However, financing is also provided by the American company producing the coal gondolas. In total, 40 train sets (each consisting of 20 rail cars) will be deployed by the utility and are intended to replace an equal number of aged steel-based train sets that will be taken out of service regardless of which type of new rail cars will be purchased.

PROJECT ADDITIONALITY:
Aluminum freight cars are used widely in North America and Europe, and have become a commercially viable option in these regions of the world. So far, only a limited number of aluminum coal gondolas have been sold on the Chinese market, all of which have been manufactured abroad. The project is still a free rider. Although, aluminum cars are more expensive than steel cars, the additional up-front cost can be recovered within two years through the significant fuel savings. Thus, the utility would have invested in the aluminum cars even without the income generated through the potential sale of carbon offset program emission reduction credits. The project is therefore not additional.

PROJECT EMISSIONS:
It is estimated that each new train set will use diesel at a rate of 0.00041 gallon/gross ton mile. The emission factor for diesel fuel is 22.384 lbs CO_2/gallon. Therefore, the project's emission rate (ER) can be estimated as follows:

$$ER = (0.00041 \text{ gallon/ton mile}) (22.384 \text{ lbs } CO_2/\text{gallon})$$

$$ER = 0.0092 \text{ lbs } CO_2/\text{ton mile}$$

PROJECT BENCHMARKS:
For the analysis of this project, we assume that 423 train sets have been deployed in China during the past 5 years. Of these, 36 are made of aluminum. Estimated fuel use and emission rate for these train sets are as follows:

Rail Transport Entity	Number of Train Sets	Type	Fuel Economy (gallon/ton mile)	Emission Factor - Diesel (lbs CO_2/gallon)	Emission Rate of Train Sets (lbs CO_2/ton mile)
Entity 1	10	Alum	0.00040	22.384	0.0090
Entity 2	26	Alum	0.00043	22.384	0.0096
Weighted Average (Alum)					0.0094
Entity 3	45	Steel	0.00052	22.384	0.0116
Entity 4	25	Steel	0.00054	22.384	0.0121
Entity 5	50	Steel	0.00051	22.384	0.0114
Entity 6	30	Steel	0.00053	22.384	0.0119
Entity 7	40	Steel	0.00055	22.384	0.0123
Entity 8	35	Steel	0.00053	22.384	0.0119
Entity 9	20	Steel	0.00054	22.384	0.0121
Entity 10	55	Steel	0.00056	22.384	0.0125
Entity 11	60	Steel	0.00052	22.384	0.0116
Entity 12	27	Steel	0.00051	22.384	0.0114
Weighted Average (Steel)					**0.0119**
Weighted Average (Sector)					**0.0117**

PROJECT ANALYSIS TABLE: Project Number TS3

	U.S. Proposal	EU Proposal	Full Technology Matrix	Hybrid Technology Matrix
Does project qualify?	If the percentile threshold test for the electricity and industrial sectors is applied, then data required to establish a percentile threshold is unavailable. The threshold test should be based on regional data.	Project qualifies as additional under the EU's positive list, because it involves more efficient and less polluting modes of transportation, and improves or substitutes existing vehicles.	The project deploys an advanced technology that is not yet fully commercial or penetrated the market in China. Thus, the project will qualify as additional.	The project qualifies as additional under the technology matrix. Thus, it also qualifies as additional under the hybrid technology matrix.
Is the project correctly identified as either a free rider or an additional project?	Indeterminate. Only two data points are given for the aluminum rail cars; thus, a percentile distribution cannot be established.	No. The project meets the criteria of the positive list, but the utility would have invested in the project even without the income generated through the sale of credits.	No. The project is additional under the technology matrix criteria, but the project would have occurred without the income generated through the sale of credits.	No. The project is additional under the technology matrix criteria, but the project would have occurred without the income generated through the sale of credits.
Number of credits Awarded	The U.S. approach does not clarify whether credits are awarded by comparing the project to a sector average, or the activity to be replaced. This is a replacement project, so if the project qualifies for credits, then credits awarded would be determined by subtracting the project ER from the ER of the steel rail cars to be replaced. However, information is unavailable to calculate the ER of the cars to be replaced.	Not applicable.	Under the technology matrix, credits for projects designed to replace existing vehicles rather than meet new demand should use the project-specific approach for baseline development, if the emissions of the vehicles to be replaced are readily identifiable. In this case, the ER of the vehicles to be replaced is unknown, thus the number of credits awarded to this project is indeterminate.	This is a replacement project, so if the project qualifies for credits, then credits awarded would be determined by subtracting the project ER from the ER of the steel rail cars to be replaced. However, information is unavailable to calculate the ER of the cars to be replaced.
Error in credits Awarded	Unknown.	Not applicable.	Project is a free rider; thus, error is equal to 100 percent of the credits awarded. In this case, the value is unknown	Project is a free rider; thus, error is equal to 100 percent of the credits awarded. In this case, the value is unknown.

METHODOLOGY ASSESSMENT: This free-rider project fails to be correctly identified as such under any of the four approaches. Under the U.S. approach, if the percentile threshold test for the electricity and industrial sectors is applied, then data required to establish a percentile threshold is unavailable. In order to establish a percentile distribution for this project, we must look at data for aluminum rail cars, as under the U.S. approach, the percentile distribution test must be based on "recent and comparable activities." Because it is proposed to replace the steel cars with aluminum cars, the percentile distribution is based on data given for "recent and comparable" aluminum rail cars. Only two data points are given for the aluminum rail cars; thus, a percentile distribution cannot be established. Instead, the threshold test should be based on regional data. This particular transportation example illustrates the problem of the U.S. approach in determining additionality when data is limited. In this case, because only two data points were available, a percentile distribution test for additionality was impossible, and additionality of the project could not be determined.

For the process of estimating emissions credits, this particular project scenario represents a situation in which it is difficult to determine whether the project is or should be treated similarly to replacement projects in the power generation sector. As the project description indicates the new aluminum rail cars will replace an equal number of aged steel-based train sets. However, it is indicated that these steel cars will be removed from service regardless of the type of cars that are chosen to replace them. Under the U.S. methodology, it is recommended that replacement projects in the power sector should be awarded credits by subtracting the project emissions from the emissions of the coal plant that the retrofit project would be replacing. However, many transportation replacement projects involve replacing vehicles or other technologies that would have been taken out of service, regardless of the carbon offset program, because they have reached the end of their life cycle. For this type of project, it is inappropriate to compare the project emission rate with the emission rate of the technology to be taken out of service. The credits awarded would be significantly inflated in comparison to a scenario in which the project emission rate was compared to a sector average of newer, comparable vehicles. Instead, the project should be compared with the alternative conventional new transportation technology that most likely would have been purchased without the prospect of carbon offset program participation.

This same distinction should be made for projects evaluated under the technology matrix approaches. Currently, the guidance for transportation sector replacement projects merely states that "projects designed to replace existing vehicles rather than meet new demand should use the project-specific approach for baseline development, if the emissions of the vehicles to be replaced are readily available." Presumably projects, where adequate emissions data is available for the technology to be replaced, will result in a pretty accurate determination of potential credits due to the use of the project-specific approach. However, in those cases where information on the old vehicles is unavailable, the guidance should be expanded to specify how projects, involving the retirement of vehicles/technologies before the end of their life cycle, should estimate the credits to be awarded.

Moreover, for both the U.S. and technology matrix approaches, we recommend that for replacement transportation projects (such as the one examined in this case study) an average of recently manufactured technologies, similar to the one used in the project, should be used to calculate the credits, if data/information on the emissions and type of transportation technology intended to replace the old technology is available. However, if the transportation technology to replace the old technology is unknown (that is, if no data is available in the host country on the emissions performance of this technology), a sector average of all recently manufactured, comparable technologies should be used to calculate credits. In this particular case, under the U.S. and hybrid technology matrix approaches, if we were to use the sector average to calculate credits, then the credits would be awarded to the project at a rate of 0.0025 lbs CO_2/ton mile.

This case study is incorrectly identified as additional under the EU's positive list. Because the project involves more efficient and less polluting modes of transportation and improves or substitutes existing technologies, it automatically qualifies as additional under the EU's positive list. This particular project would have been conducted with or without the potential sale of emission reduction credits; thus, the project is not additional and would have occurred regardless of the carbon offset program. The EU's positive list is straightforward in its requirements for additional transportation projects. It possesses no built-in mechanism to determine whether a project would have occurred without the incentive of credits. Thus, there will always be instances when the EU positive list fails to screen out certain free rider projects such as this one.

This free rider project also fails to be identified as such under the technology matrix approaches. The project involves an advanced technology that is not yet commercial and has not yet fully penetrated the market in China. The project in question is one of the first few projects of its kind to be implemented in China. As discussed in the methodology assessment for ES4, such projects do not always come into being due to an economic subsidy such as credits. In this particular case, the project developers may have witnessed the advantages of this technology in the U.S. or other developed countries and simply decided to take the risk of implementing the technology in China. However, because it is applied on a country-by-country basis, the market penetration test will always qualify first-of-its-kind projects as additional. It would be possible to apply the market penetration test on a global, rather than country level. This would serve to tighten the market penetration test and consequently exclude free rider projects of this kind from gaining credits. This would come at a cost, however, as a tightened market penetration test may also exclude truly additional projects by ignoring the non-commercial barriers to implementation of a particular technology in other countries.

Thus, all of these approaches demonstrate the capacity to allow non-additional, free rider projects to qualify for credits. Not only does this project fail to be identified as a free rider under all approaches, but also further problems are encountered in the determination of credits to award to the project. Unless the recommended approach of using a sector average to determine credits awarded is used, then data from which to determine the number of credits to award to this project is lacking. Under all approaches, credits would be determined by subtracting the project emission rate from the emission rate of the steel rail cars to be replaced. As discussed, in this case, the emission rate of the steel rail cars is not given. In other transportation case studies, *the IPCC Guidelines for National Greenhouse Gas Inventories (1997)* was cited, and an average emission rate for a comparable type of vehicle was used as the emission rate of the current mode of transportation when it was not given in the project description. This method was effective for projects involving passenger vehicles, but in this case, the project involves rail cars, and a comparable data point is unavailable within the IPCC Guidelines. Thus, the lack of data in this particular example makes it impossible to calculate even an estimate for the credits to be awarded to this project, using the current guidance for transportation projects. Moreover, in this example, because the project is incorrectly identified as additional, the error in credits awarded is 100 percent under the full and hybrid technology matrix approaches. Thus, further guidance for alternative methods of awarding credits--possibly recommending the use of a sector average to determine credits in cases where project-specific emissions rate data is unavailable--is needed for replacement transportation projects.

PROJECT NUMBER: TS4

COUNTRY: South Africa

SECTOR: Transportation

PROJECT TITLE: Clean Diesel in Transit Buses

PROJECT DESCRIPTION: South Africa has two refineries with the capability of producing "clean diesel" fuel through gas-to-liquids (GTL) technology.[22] We assume that 35 percent of the natural gas used for making "clean diesel" is based on gas that would otherwise have been flared. Because of the rising air quality problems in Johannesburg the government has announced that emissions of sulfur, NO_x, and particulate matter from road transport should be reduced and emission standards for these pollutant have been specified for each class of vehicles in the city. As part of these standards, new transit buses deployed in the city must use a blend of low-sulfur diesel, which can only be produced by mixing regular diesel with other low-sulfur products, such as "clean diesel". However, one problem with "clean diesel" is that only certain types of the most advanced diesel engines are capable of running effectively on this fuel, raising the price of new transit buses.

A local transit authority is considering replacing its aging vehicle fleet of 75 buses with an equal number of new buses. To meet the new air quality standards, the transit manager realizes he will have to use diesel mixed with "clean diesel" to comply with the new emission standards. The manager fears that the transit authority will be unable to pay the higher price of the specialized buses without raising commuter fares. However, when the manager learns that a great part of the natural gas used for producing the new diesel blend would otherwise have been flared, he decides to sell the associated greenhouse gas emission reductions as credits under an international carbon offset program. The manager has already located a European company interested in investing in the project in exchange for the credits. In this way, the transit authority will be able to recoup some of the additional cost of using the "clean diesel".

PROJECT ADDITIONALITY: The project is a free rider. Due to the new emission standards in Johannesburg, the project will be implemented regardless of the potential sale of carbon offset program credits. However, seeing the opportunity to lower the cost

---

[22] Gas-to-liquids (GTL) technologies chemically change natural gas molecules, breaking them apart, and re-combining them with oxygen to form a mixture called synthesis gas. In turn, synthesis gas can be chemically converted into different types of hydrocarbon products like clean-burning transportation fuels (new diesel) or a variety of high-value chemicals. One of the potential uses for GTL technology and new diesel is as a replacement fuel for conventional diesel or as a blending agent with conventional fuels to help meet more stringent environmental regulations.

of the new buses, the transit authority is in effect selling the credits to the European company.

PROJECT EMISSIONS: Let us assume that the emission factor of this particular blend of clean diesel is 17.573 lbs CO_2/gallon once the carbon dioxide and methane emission reductions from utilizing the flared natural gas has been accounted for. The type of buses purchased for the transit authority has a fuel economy of 0.168 gallon/mile. Hence, the emissions rate of the project is as follows:

$$ER = (17.573 \text{ lbs } CO_2/\text{gallon}) (0.168 \text{ gallon/mile})$$

$$ER = 2.952 \text{ lbs CO2/mile}$$

PROJECT BENCHMARKS: For the purpose of this case study, we assume that 6 different transit bus operators in Johannesburg have purchased new buses within the past 2 years, reaching a total of 312 new vehicles during that period. Estimated emission rates for these diesel buses is a follows:

Transit Authority	Number of Buses	Fuel Economy (gallon/mile)	Emission Factor - Diesel (lbs CO_2/gallon)	Emission Rate of Buses (lbs CO2/mile)
Entity 1	55	0.177	22.384	3.962
Entity 2	26	0.183	22.384	4.096
Entity 3	45	0.179	22.384	4.007
Entity 4	69	0.180	22.384	4.029
Entity 5	50	0.188	22.384	4.208
Entity 6	67	0.172	22.384	3.850
Weighted Average				**4.010**

PROJECT ANALYSIS TABLE: Project Number TS4

	U.S. Proposal	EU Proposal	Full Technology Matrix	Hybrid Technology Matrix
Does project qualify?	The project will qualify even if "X"<1st percentile, or lower than 3.850 (ER<3.850).	Project qualifies as additional under the EU's positive list, because it involves more efficient and less polluting modes of transportation.	The project deploys advanced technology that is not yet fully commercial and has not fully penetrated the market in South Africa. The project will qualify as additional.	The project qualifies as additional under the technology matrix. Thus, it will also qualify under the hybrid technology matrix approach.
Is the project correctly identified as either a free rider or an additional project?	No. While the project may qualify as additional under U.S. standards, the project will be implemented regardless of the potential sale of credits. Thus, it is a free rider.	No. While the project qualifies as additional under the positive list, the project will be implemented regardless of the potential sale of credits. Thus, it is a free rider.	No. While the project qualifies as additional under the technology matrix, the project will be implemented regardless of the potential sale of credits. Thus, it is a free rider.	No. While the project qualifies as additional under the hybrid technology matrix, the project will be implemented regardless of the potential sale of credits. Thus, it is a free rider.
Number of credits Awarded	The U.S. approach does not clarify whether credits are awarded by comparing the project to a sector average or the emission factor of the activity to be replaced. In this case, the project credits are determined by subtracting the project ER from the average ER of new conventional buses (4.01 - 2.952 = 1.06 lbs CO_2/mile).	Not applicable.	In an earlier study, we recommend that credits for replacement transportation projects be derived by subtracting the ER of the project (2.952 lbs CO2/mile) from the ER of the vehicles to be replaced (3.80 lbs CO_2/mile*). Thus, credits awarded = 0.938 lbs CO_2/mile.	The project credits would be determined by subtracting the project ER from the ER of the buses/fuel to be replaced (3.89* - 2.952 = 0.938 lbs CO_2/mile).
Error in credits Awarded	Project is a free rider; thus, the error is equal to 100 percent of the credits awarded (1.06 lbs CO_2/mile).	Not applicable.	Project is a free rider; thus, error is equal to 100 percent of the credits awarded (0.938 lbs CO_2/mile.	Project is a free rider; thus, the error is equal to 100 percent of the credits awarded (0.938 lbs CO_2/mile).

147

METHODOLOGY ASSESSMENT: This free rider project qualifies for credits under all four approaches. As mentioned previously, the U.S. approach does not provide guidance on benchmark development for the transportation sector. If we apply the "recent and comparable" additionality test, the project will qualify for credits even if the threshold is set at the most stringent level or the 1st percentile (ER<3.850). Because this is a replacement project involving the retirement of vehicles that have reached the end of their life cycle, we recommend that the estimation of credits will be based on a comparison of project emissions with the emissions of recent and comparable vehicles (see TS3). In this case, project emissions were compared to the weighted average of new, conventional diesel buses purchased within the last two years. Accordingly, the U.S. approach awarded this project credits at a rate of 1.06 lbs CO_2 per mile. As previously noted, the U.S. methodology should be clarified to provide more detailed guidance on the estimation of credits from replacement projects versus new capacity projects.

Under the U.S. approach, as discussed in the methodology assessment for IS4, a certain number of free rider projects will qualify for credits, even if the percentile test is set at the most stringent level. By simply requiring emission reduction projects to be "significantly better than average," the U.S. percentile test guarantees that a certain number of business-as-usual projects will qualify for credits. The comparison of the project emission rate with a benchmark value does not directly address the issue of free ridership; hence, the test may fail even when "X" is set at a very stringent level.

Within the transportation sector, many existing laws and regulations require compliance with air quality standards, thus rendering projects such as this one as projects that would occur regardless of the potential of earning credits. This is quite problematic, as many free rider replacement transportation projects would qualify for credits because the project vehicles are more efficient or emit less CO2 than the older vehicles to be replaced. The issue of parallel regulations mandating improved technology deployment appears particularly dominant within the transportation sector compared to other sectors, such as the power generation and industrial sectors. Thus, it may be useful to add a requirement that any technology already required by other regulations should be excluded from transportation sector projects.

As transportation projects that improve efficiency or substitute existing vehicles are included on the EU's positive list, this project also qualifies as additional under this approach. As discussed in the previous example, the EU's positive list is straightforward in its requirements for determining additionality, and has no built-in mechanism to determine whether a project would have occurred without the incentive of credits. Thus, there will always be instances when the EU positive list fails to screen out certain free rider projects such as this one, which would have occurred even without the incentive of credits.

The two technology matrix approaches also qualify this free rider project for credits. This project involves an advanced technology that is not yet commercial and has not yet fully penetrated the market in South Africa. As discussed in the previous example, the project in question is one of the first few projects of its kind to be implemented in South

Africa. As discussed, such projects do not always come into being due to an economic subsidy such as credits. Further, because it is applied on a country-by-country basis, the market penetration test will always qualify first-of-its-kind projects as additional. Thus, the technology matrix approach, like the U.S. and EU approaches, has the potential to incorrectly classify free rider projects as additional.

In this case, the technology matrix does not award credits at the same rate as the U.S. and hybrid technology matrix approaches--0.938 lbs CO_2 per mile. Following the guidance of a previous report (Developing the Technology Matrix for India and Ukraine) we applied the project-specific approach to estimate the emission credits and compared the emissions of the new vehicles to that of the old retired buses. However, this analysis does not distinguish between replaced vehicles that have not yet reached the end of their life cycle and the ones that are obviously ready for retirement. For this particular type of project, where the old buses will be taken out of service no matter what, this distinction should be made to avoid inflating the credits awarded to the project. The guidance for the technology matrix should be expanded to specify that projects retiring vehicles that still have not reached the end of their life cycle should compare project emissions to the emissions of the old vehicles. Mean while projects involving the replacement of completely aged vehicles should compare project emissions to the emissions of recent models of comparable conventional technologies.

PROJECT NUMBER: TS5

COUNTRY: Mexico

SECTOR: Transportation

PROJECT TITLE: Electric Vehicles in Mexico City

PROJECT DESCRIPTION: Mexico City is one of the world's most polluted cities. A major part of the city's air quality problems is caused by the transportation sector. To reduce pollution and greenhouse gas emissions from road transport, the government has identified the deployment of electric vehicles as an alternative to vehicles using internal combustion engines.[23]

This project involves the purchase and deployment of 125 electric passenger vehicles to replace a similar number of aging gasoline internal combustion vehicles with an average age of 7 years. The vehicles are purchased by the city government to be used by city workers for business-related transportation. The city's fleet parking area will be installed with the appropriate recharging facilities and the mechanics servicing the vehicles will receive special training on the maintenance and repair of electric vehicles. The American firm producing the electric vehicles is helping to finance the project in exchange for the credits accrued during the life of the project.

PROJECT ADDITIONALITY: The project is additional. The purchase and deployment of the electric vehicles and related infrastructure would not have been undertaken without the favorable investment terms provided by the American vehicle manufacturer. The American manufacturer provided the favorable financing specifically with the goal of obtaining credits.

PROJECT EMISSIONS: The batteries used for charging the electric vehicles are assumed to run on electricity generated from the average mix of generating capacity in Mexico. The estimated emission rate of the electric vehicles is 23.98 g CO_2/mile.[24] The estimated emission rate of the gasoline internal combustion vehicles that would have been purchased absent the project is 34.36 g CO_2/mile.

$$ER = 23.98 \text{ g } CO_2/\text{mile}$$

PROJECT BENCHMARKS: In this project, we assume that "recent and comparable" activities mean new gasoline internal combustion vehicles sold in Mexico in 1999. During that year, we assume 12 passenger vehicle models were sold on the Mexican market. In total, 365,421 gasoline internal combustion vehicles were sold in 1999. The emission rates for these vehicles are as follows:

---

[23] National Climate Change Action Plans: Report for Developing and Transition Countries, Mexico. U.S. Country Studies Program. Washington, D.C., October 1997.
[24] Ibid.

Emission Factors for Gasoline Internal Combustion Passenger Vehicles Sold in Mexico, 1999			
Control Technology	**Number of Vehicles Sold**[25]	**Share of Total Vehicles Sold (%)**	**CO_2 Emissions (g/mile)**
Model 1	12,802	3.22	35.63
Model 2	6,308	1.59	34.78
Model 3	538	0.14	34.12
Model 4	7,245	1.83	36.33
Model 5	68,448	17.24	35.97
Model 6	3,542	0.90	32.88
Model 7	233,642	58.86	35.09
Model 8	4,562	1.15	34.44
Model 9	379	0.10	33.11
Model 10	48,345	12.18	33.64
Model 11	9,214	2.32	36.02
Model 12	1,943	0.49	35.26
Weighted Average			**35.09**
Total	**396,968**	**100**	

---

[25] 1999 Ward's Automotive Yearbook

PROJECT ANALYSIS TABLE: Project Number TS5

	U.S. Proposal	EU Proposal	Full Technology Matrix	Hybrid Technology Matrix
Does project qualify?	The project will qualify even if "X"<1st percentile, or lower than 32.88 (ER<32.88).	The project qualifies as additional under the EU's positive list because it involves more efficient and less polluting modes of transportation, and improves or substitutes existing vehicles.	The project deploys an advanced technology that is not yet fully commercial and does not exhibit market penetration in Mexico. Thus, the project will qualify as additional.	The project qualifies as additional under the technology matrix; thus, it will also qualify as additional under the hybrid technology matrix.
Is the project correctly identified as either a free rider or an additional project?	Yes.	Yes.	Yes.	Yes.
Number of credits Awarded	The U.S. approach does not clarify whether credits for replacement projects are awarded by comparing the project to a sector average or to the ER of the activity to be replaced. In this case, we subtract the ER of the project from the ER of the vehicles to be replaced; thus, credits awarded will be 34.36 - 23.98 = 10.4 g CO_2/mile.	Not applicable.	credits for replacement transportation projects are derived by subtracting the ER of the project (23.98) from the ER of the vehicles to be replaced (34.36). Thus, credits awarded = 10.4 g CO_2/mile.	The project credits would be determined by subtracting the project ER from the ER of the vehicles to be replaced. Thus, credits = 34.36 - 23.98 = 10.4 g CO_2/mile.
Error in credits Awarded	Unknown.	Not applicable.	Unknown.	Unknown.

METHODOLOGY ASSESSMENT: This additional project qualified as such under all approaches. As previously discussed, the U.S. approach does not provide guidance on benchmark development for the transportation sector. However, according to the U.S. approach for other sectors, the additionality test involves a comparison with a reference scenario consisting of a set of "recent and comparable" activities. In this case, it is assumed that "recent and comparable" activities mean new gasoline internal combustion vehicles sold in Mexico in 1999. Applying the U.S. approach to this set of data, this electric vehicle project will qualify as additional regardless of the percentile threshold selected.

This project is a replacement project that involves a switch from 125 aging gasoline internal combustion vehicles to the same number of electric passenger vehicles. As previously addressed, the U.S. guidance on estimating credits from replacement versus new capacity projects is unclear and should be clarified. According to the U.S. methodology, credits are awarded to replacement projects in the power sector by subtracting the project emissions from the emissions of the activity to be replaced. Because this is a replacement project involving the retirement of vehicles that have not yet reached the end of their life cycle, it is recommended that the emissions rate of the project is subtracted from the emissions rate of the vehicles to be replaced. In this case, the emissions rate of the vehicles to be replaced was given--34.36 g CO_2/mile. Accordingly, the U.S. approach awards credits at a rate of 10.4 g CO_2/mile. However, as previously discussed, the U.S. methodology should be clarified to provide more detailed guidance on how to estimate credits from replacement projects versus new capacity projects.

This project also qualifies as additional under the EU's positive list, as this is a transportation project that improves efficiency or substitutes existing vehicles. Both the full and hybrid technology matrix approaches also correctly identify the project as additional. In this case, the technology matrix also awards credits at a rate of 10.4 g CO_2/mile.

This example illustrates the fact that like the CNG replacement vehicle projects highlighted in TS1, electric vehicle replacement projects appear to represent a class of projects that are handled equally well by all four approaches. The four baseline methods correctly identify the project as additional and the credits awarded are the same for the U.S. and technology matrix approaches.

COUNTRY: Thailand

SECTOR: Transportation

PROJECT TITLE: Smart Toll System

PROJECT DESCRIPTION: In Thailand, passenger demand is estimated to grow at a rate of 7.8 percent per year during 2000 to 2010.[26] Freight demand is expected to grow at a rate of 3.6 percent per year during that same period. A large part of this growth is concentrated around the city of Bangkok, which has witnessed a remarkable population growth caused by migration and economic expansion. As a result, the city is troubled by traffic congestion and bad air quality. Better urban planning will improve system-wide transportation efficiency by minimizing travel time and congestion.

This project involves the deployment of a smart toll system on a major highway that is under construction to connect three heavily populated industrial and residential areas in the suburbs. This is the first electronic toll system to be built in the country. The smart toll system will include the construction of 16 smart toll points along the highway. The electronic toll system can only be used by passenger vehicles and light-duty trucks. Heavy-duty vehicles will have to use manual toll booths. By allowing commuters to use electronic charge cards for passing through the toll areas, city planners expect to reduce fuel use by 16.8 percent on this particular stretch of highway through considerable reductions in waiting times at toll points. As such, the project is a small component of a larger effort to improve traffic management and reduce congestion. However, the smart toll project has been singled out for participation in an international carbon offset program because the city government is more confident that the emission reductions from this part of their traffic management activities can be quantified and verified. Moreover, the deployment of the smart toll system is undertaken specifically with the purpose of reducing congestion, fuel use, and thus greenhouse gas emissions. The electronic toll system is significantly more costly than regular toll booths. However, the city government has been able to improve the economics of the project by selling the rights to the future emission reductions to an American investor.

PROJECT ADDITIONALITY: Even though electronic toll systems are commercially available in North America, Europe and Asia, the smart toll system would not have been implemented without the possibility of selling the expected credits. Instead, a regular toll system would have been built along the highway. Moreover, the American investor is participating in the project solely with the purpose of obtaining credits. Hence, the project is additional.

---

[26] National Climate Change Action Plans: Interim Report for Developing and Transition Countries. Thailand. U.S. Country Studies Program. Washington D.C., October 1997.

PROJECT EMISSIONS: The city government estimates that without the electronic toll collection system no more than 600 vehicles per hour could pass through a similar-sized manual toll system. In comparison, it is estimated that the smart toll system will be able to handle up to 2,100 vehicles per hour, a capacity increase of 350%. This will lead to an estimated reduction of fuel consumption of 16.8%, or a reduction of 24,499,448-vehicle km a year. The default emission factor per vehicle km in Bangkok is 531 g CO_2 equivalent per km.

PROJECT BENCHMARKS: There are no available statistics on the average reduction of vehicle km traveled that can be achieved by installing electronic toll collection systems. However, an analysis by the U.S. government found that if a toll collection capacity increased by at least 250% by installing electronic lanes, fuel consumption would decrease by up to 12%.

PROJECT ANALYSIS TABLE: Project Number TS6

	U.S. Proposal	EU Proposal	Full Technology Matrix	Hybrid Technology Matrix
Does project qualify?	Data is insufficient to perform the percentile test and establish a threshold; therefore, it cannot be determined whether or not the project would qualify.	The positive list allows for improvements in transport energy consumption. The project will result in fuel consumption reductions and will qualify.	The project involves the use of a commercially available technology. However, the technology has never before been used in the country; therefore, the project qualifies under the technology matrix.	The project involves the use of a commercially available technology. However, the technology has never before been used in the country; therefore, the project qualifies under the technology matrix.
Is the project correctly identified as either a free rider or an additional project?	Indeterminate. Threshold cannot be established.	Yes	Yes	Yes
Number of credits Awarded	Indeterminate. This is an energy consumption project and the U.S. proposal has not been developed for this type of project. A baseline could be developed by determining what the average reductions in fuel consumption would be.	Not Applicable	Under project specific credits = .000531 metric tons CO_2/km X 24,499,448 vehicle km/yr = 13,009.2 metric tons CO_2/vehicle km/yr	Under project specific credits = .000531 metric tons CO_2/km X 24,499,448 vehicle km/yr = 13,009.2 metric tons CO_2/vehicle km/yr
Error in credits Awarded	Unknown	Not Applicable	Unknown.	Unknown.

METHODOLOGY ASSESSMENT: This additional project is identified as such only by the EU proposal. The project would not qualify under the technology matrix approach but may qualify under the fall-back project specific approach. Under the U.S. proposal, the project is indeterminate.

This project is unique in the sense that it does not deal directly with vehicles but rather with improvements to the transportation infrastructure. Although the project would qualify under the EU proposal, it is a type of project that is problematic for all of the proposals. The problem stems from the fact that quantifying the emission reductions resulting from the project would be difficult. Under the U.S. proposal, this problem is magnified because the percentile test requires a comparison of emission rates of recent and comparable facilities. In the absence of data to establish a threshold, the U.S. proposal cannot determine whether or not the project would qualify. Moreover, identifying comparable facilities for comparison may prove to be very difficult as there are no other smart toll systems operating in the host country and the transportation infrastructure of other countries within the host country's region is likely to differ significantly in terms of design and vehicle use. The only possible indicator would be a comparison of the percentage reduction of vehicle miles traveled for each electronic toll system in the region.

More importantly however, the U.S. proposal does not provide guidance on the transportation sector. Further, because this is an energy consumption improvement project it is not clear that the U.S. proposal would accommodate this type of project. Although it would reduce emissions, the project will not affect the emission rates of the vehicles using the smart toll system. By allowing faster vehicle speed on the highway, the project improves the utilization of the vehicles, not the efficiency of their energy use. If the U.S. proposal specifies the transportation sector percentile test to be based on energy efficiency or emission rates, as it does for the industrial sector, all transportation infrastructure improvement projects that result in energy consumption reductions cannot be subjected to the threshold test. Indeed, the smart toll project indicates that there are several project types in the transportation sector that cannot be analyzed via the U.S. standardized approach as it is currently designed. This reiterates the need for a backup approach and detailed guidelines for when to apply this backup. These guidelines could include a list of project types that should apply the backup methodology and those project types that could use the standardized approach.

As for the technology matrix, the smart toll technology is commercially available so the project would not qualify if only the economics of the technology are examined. However, this project is the first-of-its-kind in Thailand. Because it has no market penetration in the country, the project will qualify in the technology matrix. Emission reduction calculations would be derived by estimating the reduction in vehicle km traveled per year and multiplying this information by the emission rate per vehicle mile traveled (credits = .000531 metric tons CO_2/km x 24,499,448 vehicle km/year = 13,009.2 metric tons CO_2/vehicle km/year).

PROJECT NUMBER: TS7

COUNTRY: Ukraine

SECTOR: Transportation

PROJECT TITLE: 46 New Conventional Diesel Buses

PROJECT DESCRIPTION: In Ukraine, economic growth has slowed during the last 10 years putting a halt to industrial development and expansion. In spite of the general economic slowdown, the transportation sector has continued to grow and demand for passenger transport has risen significantly. However, due to the poor economic health of the country, the government has been unable to maintain and adjust the country's public transportation system to the meet the growing demand for transport. In particular, the bus system is deteriorating as existing vehicles age and are kept in service due to the lack of capital to invest in new buses. Specifically, buses are over-crowded and scheduled bus routes are often delayed due to chronic maintenance problems and lace of bus availability. Due to the lack of an efficient public transportation system, the purchase of passenger vehicles is growing even faster.

This project involves the purchase of 46 new conventional diesel transit buses by the transit authority servicing the central area of Kiev. The buses will be deployed to meet new demand for transportation. No old vehicles will be taken out of service. The purpose of the project is to reduce the use of personal transportation and thus decrease fuel use. As access to capital for purchasing new buses is limited, the local transit authority has been looking for a foreign investor to finance part of the project in return for the rights to the potential emission reduction credits from the project. An American energy company has expressed interest in purchasing the credits from the project.

PROJECT ADDITIONALITY: The project is additional. The project is undertaken specifically with the objective of reducing greenhouse gas emissions through reduced use of personal passenger travel. Moreover, the local transit authority would not have considered purchasing the new buses without the additional investment obtained through the sale of credits.

PROJECT EMISSIONS: The CO_2 exhaust emission factor of the new diesel buses is 987 g CO_2/km. The CH_4 emission factor is 0.04 g CH_4/km. The GWP of CH_4 is 21. The project's emission rate is as follows:

ER for CO_2 = 987 g CO_2/km

ER for CH_4 = (0.04 g CH_4/km)(21) = 0.84 g CO_2 equivalent/km

ER project = 987 g CO_2/km + 0.84 g CO_2 equivalent/km = 987.84g CO_2 equivalent/km

PROJECT BENCHMARKS: The average passenger car in Ukraine is assumed to drive 4.7 km/liter.[27] The average emission factor for passenger vehicles in Ukraine is assumed to be 0.14 g CH4/km and 506 g CO2/km.

We assume that four different transit bus operators in Kiev have purchased new buses (all diesel fueled) within the past 3 years, reaching a total of 85 new vehicles during that period. Estimated emission rates for these diesel buses is a follows:

Transit Authority	Number of Buses	Fuel Economy (km/liter)	Emission Factor - Diesel (g CO_2/litre)	Emission Rate of Buses (g CH4/km)	Emission Rate of Buses (g CO2/km)
Entity 1	15	2.3	22.384	0.07	1017
Entity 2	22	2.1	22.384	0.13	1046
Entity 3	30	2.2	22.384	0.11	1038
Entity 4	18	2.3	22.384	0.07	1017
Weighted Average				**0.10**	**1032**

---

[27] This average fuel economy data is derived from the 1996 Revised IPCC National Greenhouse Gas Inventories: Reference Manual. According to the IPCC the vehicle category for most developing countries is "uncontrolled" vehicle technology. We assume that this category will also apply to vehicles in Ukraine.

PROJECT ANALYSIS TABLE: Project Number TS7

	U.S. Proposal	EU Proposal	Full Technology Matrix	Hybrid Technology Matrix
Does project qualify?	The project will qualify even if "X"<1st percentile, or lower than 1017 (ER<1017).	Under the positive list, projects that involve "more efficient and less polluting modes of transportation" will qualify. The buses in this project have a better ER compared to the average of the reference scenario; thus, the project will qualify as additional.	Project does not involve advanced, non-commercial technology; thus, it does not qualify as additional. Similar vehicles are already being sold on the Ukrainian market. Project developers may still attempt to prove additionality using project-specific approach.	This project does not qualify as additional under the technology matrix. Thus, it will not qualify under the hybrid technology matrix approach. Project developers may still attempt to prove additionality using the project-specific approach.
Is the project correctly identified as either a free rider or an additional project?	Yes.	Yes.	No. Project developers may attempt to prove additionality by proving that a lack of funding for the project exists; however, this would be very difficult.	No. Project developers may attempt to prove additionality by proving that a lack of funding for the project exists; however, this would be very difficult.
Number of Credits Awarded	Credits are determined by comparing project ER to the average ER of the recent, comparable buses or by comparing project ER to passenger miles to be replaced by bus use. If the former, then we use average ER of new diesel fueled buses; thus, Credits = 1032 - 987.84 = 44.16 g CO_2/km. If the latter, further data is necessary to quantify the ER of the individual passenger vehicles/miles that would be replaced via bus use.	Not applicable.	Project does not qualify for credits.	Project does not qualify for credits.
Error in Credits Awarded	Unknown.	Not applicable.	Project is not awarded credits, but needs credits to be economically feasible; thus, it will not be undertaken.	Project is not awarded credits, but needs credits to be economically feasible; thus, it will not be undertaken.

METHODOLOGY ASSESSMENT: This additional project qualifies for credits under the U.S. and EU approaches. If the "recent and comparable" additionality test is applied to this case study, then the project will qualify for credit even if the threshold is set at the most stringent level (ER<1,017), and credits are awarded at a rate of 44.16 g CO_2 per km. This determination was made by comparing the project emissions rate to the average emissions rate of new diesel-fueled buses purchased in Kiev in the past three years. Thus, similar to case study TS2, this case study illustrates that a vehicle model which is already available on the Ukrainian transportation market is able to receive credits under the U.S. baseline approach. However, for this new capacity project under the U.S. approach, credits could also be calculated by comparing the project emissions rate to the passenger miles that would be replaced by bus use. This would prove to be a difficult task, however, because a study must be conducted to determine the number of individuals that would be expected to ride the new buses and the amount of miles to be traveled by these individuals in order to estimate an average emissions rate of passenger vehicles to be replaced by the buses. Currently, information is unavailable to make such a calculation. For this type of project, the U.S. approach should provide guidance on which alternative approach to use, such as the project-specific approach.

This type of transportation project is also classified as additional under the EU's positive list. In the EU's proposed positive list of technologies, it is stated that "more efficient and less polluting modes of mass and public transport" will qualify for credit. The buses included in this project have a better emission rate compared to the average of the reference scenario; thus, the project involves "less polluting" modes of transportation and will qualify.

The two technology matrix approaches disqualify this project because it applies conventional vehicle technology that has already been introduced in Ukraine. As such, the matrix approaches fail to qualify a legitimately additional project for credits. Although the project fails to qualify under the full and hybrid technology matrix approaches, the project developers could attempt to demonstrate project additionality using the project-specific approach. In such a case, project developers would attempt to prove additionality by confirming that a lack of funding for the project exists. This, however, would prove to be quite difficult. Thus, some truly legitimately additional projects, which are regular capacity expansions, will be denied qualification for credits under the technology matrix approaches.

The implications of the failure of this project to qualify for credits are that the project would not be undertaken, and citizens of the Ukraine would presumably continue to purchase new passenger vehicles. This project, involving the deployment of the new buses to meet new demand for transportation and reduce the use of personal transportation (thereby decreasing fuel use), is truly additional, and as such, requires the incentive of credits in order to be deemed feasible. The failure to qualify truly additional projects such as this one will reduce the amount of credits awarded to the U.S. and other developed countries, thereby raising the costs of implementation.

PROJECT NUMBER: TS8

COUNTRY: India

SECTOR: Transportation

PROJECT TITLE: New Two-Wheelers

PROJECT DESCRIPTION: In India, the fastest growing vehicle type is two-wheelers. Thus, efforts to reduce emissions from two-wheelers will have a significant impact on urban air quality and overall greenhouse gas emissions from the transportation sector. Because no environmental control regulations were in place for two-wheelers before 1991, vehicles put into service before that year provide a disproportionately much larger share of emissions than more recent models.

This project involves the purchase of 275 new highly fuel-efficient four-stroke engine two-wheelers by the Federal Government housed in New Delhi. The two-wheelers, powered by a mix of gasoline and oil lubricants, will retire and replace an equal number of old two-stroke engine two-wheelers that were put into service before 1991. These vehicles were targeted for retirement because an inspection of the entire government vehicle fleet identified them as ill-maintained and seriously lacking in emission control technology. Thus, the project will lead to emission reductions both through the upgrade to newer, more efficient vehicles as well as through the switch to the more fuel-efficient four-stroke engines. Realizing that the project will lead to significant greenhouse gas emission reductions the project managers have approached a European energy company and offered to sell the potential credits in return for investment in the project.

PROJECT ADDITIONALITY: The project is not a free rider. The Federal Government would not have purchased this particular type of highly efficient two-wheelers without the potential of participating in an international carbon offset program. Although the new two-wheelers are using a technology that is already available in India, they are more expensive than regular two-wheelers. The Federal government only decided to purchase the most efficient two-wheelers because of the potential for obtaining additional financing through the sale of credits. The European investor is sponsoring the project solely with the objective of purchasing credits.

PROJECT EMISSIONS: The CO_2 exhaust emission factor of the new four-stroke engine two-wheelers is 266 g CO_2/km.[28] The CH_4 emission factor is 0.26 g CH_4/km. The global warming potential (GWP) of CH_4 is 21.[29] The project's emission rate is as follows:

---

[28] These emission factors only include exhaust emissions. GHG emissions during the production, transportation, and distribution of the fuels are not included in this equation.

[29] Intergovernmental Panel on Climate Change (IPCC), 1997. "Revised 1996 IPCC Guidelines for National Greenhouse Gas Inventories"

ER for CO_2 = 272 g CO_2/km

ER for CH_4 = (0.26 g CH_4/km)(21) = 5.46 g CO_2 equivalent/km

ER project = 272 g CO_2/km + 5.46 g CO_2 equivalent/km = 277.46 g CO_2 equivalent/km

The average emission rate of the old vehicles to be replaced is 346.12 g CO_2 equivalent/km.

PROJECT BENCHMARKS: No data is available describing exhaust emissions of carbon dioxide and methane from different types of road transportation vehicles in India. Instead, we have used fictional average emission factors for the different models of new four-stroke engine two-wheelers sold in India in 1998. All of the vehicle models included are powered by a mix of gasoline and oil lubricants.

Estimated Emission Factors for New Four-Stroke Engine Two-Wheelers Sold in India, 1998			
Control Technology	Number of Two-Wheelers Sold	Share of Total Two-Wheelers Sold (%)	CO2 Equivalent (g/km)
Model 1	19,208	6.0	285.5
Model 2	13,573	4.3	279.8
Model 3	7,483	2.4	278.5
Model 4	18,258	5.7	281.6
Model 5	58,448	18.3	282.8
Model 6	3,542	1.1	275.8
Model 7	113,672	35.5	283.8
Model 8	21,349	6.7	280.6
Model 9	17,552	5.5	282.8
Model 10	46,392	14.5	278.8
Average	**N/A**	**N/A**	**281.0**
Weighted Average	**N/A**	**N/A**	**282.2**
Total	**319,477**	**100**	**N/A**

PROJECT ANALYSIS TABLE: Project Number TS8

	U.S. Proposal	EU Proposal	Full Technology Matrix	Hybrid Technology Matrix
Does project qualify?	If $X < 10^{th}$ percentile, then the project will not qualify (ER > 275.8). If 10th percentile $< X < 15^{th}$ percentile, then the project will qualify ($275.8 < X < 278.5$).	The project qualifies as additional under the EU's positive list because it involves more efficient and less polluting modes of transportation, and improves or substitutes existing vehicles.	Project does not deploy an advanced technology. It uses a technology already available in India; thus, it is not considered additional under the technology matrix. Project developers may use the project-specific approach. Given that the new technology is expensive and will only be used if financing is available through credits, this project will likely qualify as additional.	Project developers may use the project-specific approach, and given that the new technology is expensive and will only be used if additional financing is available through credits, this project will likely qualify as additional.
Is the project correctly identified as either a free rider or an additional project?	Yes, if 10th percentile $< X < 15^{th}$ percentile.	Yes.	Yes, while the technology is already used in India, project developers may use the project-specific approach to show that the project would not be undertaken without the potential credits.	Yes, project developers may use the project-specific approach to show that project would not be undertaken without the potential credits.
Number of credits Awarded	If $X < 10^{th}$ percentile, then no credits will be awarded to this project. If 10th percentile $< X < 15^{th}$ percentile, then the project will qualify for credits; however, the U.S. approach does not clarify whether credits for replacement projects are awarded by comparing the project to a sector average or to the ER of the activity to be replaced. In this case, we subtract the ER of the project from the ER of the vehicles to be replaced (credits = $346.12 - 277.46 = 68.66$ g CO_2/km).	Not applicable.	credits for replacement transportation projects are derived by subtracting the ER of the project (277.46) from the ER of the vehicles to be replaced (346.12). Thus, credits awarded = 68.66 g CO_2/km.	The project credits would be determined by subtracting the ER of the project from the ER of the vehicles to be replaced (credits = $346.12 - 277.46 = 68.66$ g CO_2/km).
Error in credits Awarded	Unknown.	Not applicable.	Unknown.	Unknown.

METHODOLOGY ASSESSMENT: This additional project qualified as such under all approaches, except for in a certain scenario under the U.S. approach. If the U.S. eligibility threshold test for additionality is applied to this transportation sector project, then the project will qualify as additional only if "X" is set between the 10th and 15th percentile. However, the project will not qualify as additional if "X" is less than the 10th percentile. Because the U.S. approach does not offer guidance on the value that must be chosen for "X," the result of this additionality determination is indeterminate. If we assume that "X" may be set between the 10th and 15th percentile, then the project will qualify, and credits will be awarded to the project. As stated, however, the U.S. should fully clarify an approach for determining additionality and clearly indicate a level to which "X" must be compared in order to avoid indeterminate additionality qualifications such as this one.

This project involves the replacement of 275 ill-maintained and inefficient, old two-stroke engine two-wheelers that were put into service before 1991. The project replaces these vehicles with 275 new, highly fuel efficient, four-stroke engine two-wheelers. As previously discussed, the U.S. guidance on the estimation of credits for replacement projects versus new capacity projects is unclear and should be clarified. According to the U.S. methodology, credits are awarded to replacement projects in the power sector by subtracting the project emissions from the emissions of the activity to be replaced. Because this project qualifies as a replacement project, we assume that credits may be determined in this same way. Thus, in this case, we subtract the emissions of the project (277.46 g CO_2/km) from the emissions of the vehicles to be replaced (346.12 g CO_2/km), and credits awarded equal 68.66 g CO_2/km. If credits were awarded to this project by comparing its emissions to a sector average, then substantially fewer (almost 15 times fewer) credits would be awarded to the project (4.74 g CO2/km). As recommended in a previous transportation sector analysis (TS4), the U.S. guidance on transportation sector projects should be expanded to distinguish between transportation projects replacing technologies that have reached the end of their life cycle and projects replacing technologies that can still be utilized. Projects replacing vehicles, such as these two-wheelers, that have not yet reached the end of their life cycle should estimate credits by comparing the project emissions to the emissions of the vehicles to be replaced. However, transportation projects that replace vehicles that would have been taken out of service anyway should be treated as new capacity expansion projects. That is, credits should be estimated by comparing the project emissions rate to that of the weighted average of new conventional vehicles.

This project clearly qualifies as additional under the EU's positive list, as this is a transportation project that improves efficiency and substitutes existing vehicles. However, under the technology matrix approaches, the project will likely only qualify if the project-specific approach is employed. Because the technology used in this project is already available and in use in India, the project technology is not considered "advanced" and "non-commercial" under the technology matrix. However, the technology matrix offers the fallback of the project-specific approach. In this case, it is highly likely that the project developers could prove additionality because the new technology is expensive and will not be utilized without the additional financial backing offered through the incentive of attaining and selling credits. This example thus illustrates that the existence of a fallback method for determining additionality is extremely important. The technology matrix is well equipped with its inclusion of the project-specific approach as a fallback method. Under both the full and hybrid technology matrix approaches, credits are awarded by subtracting the emissions of the project from the emissions of the vehicles to be replaced, thus resulting in the same value that was attained under the U.S. approach (68.66 g CO_2/km).

COUNTRY: Brazil

SECTOR: Transportation

PROJECT TITLE: Improving Road Infrastructure

PROJECT DESCRIPTION: As economic expansion continues to cause population growth and migration to urban areas, road transportation between major metropolitan areas is rising as well. This project involves the upgrade and improvement of a major highway connecting two large cities. In total, 350 miles of highway will be renovated, including placement of new asphalt and straightened highway curves. These upgrades will help improve traffic flow by allowing vehicles to move at higher speed. Particularly, rush hour bottlenecks will be avoided, thus reducing driving time and vehicle miles traveled. The project is undertaken by the state government, which is responsible for the highway. No advanced technology will be used for the project, however, it is undertaken specifically with the purpose of reducing fuel use and hence limits associated greenhouse gas emissions. An American investor has been located who is willing to purchase the greenhouse gas emission reductions to be accrued by the project. The potential for gaining additional revenue from the sale of credits helped convince the state government of the benefits of undertaking this project.

PROJECT ADDITIONALITY: The project is additional. It is undertaken specifically with the purpose of reducing greenhouse gas emissions. The availability of additional financing due to the potential sale of credits helped convince the project developers to proceed with the project.

PROJECT EMISSIONS: The improvements are estimated to reduce fuel use per vehicle mile traveled by about 5 percent.

Vehicle Type	Share of Vehicles (%)	Emission Factor (g CO_2/kg fuel)	Emission Rate (CO_2 g/km)
Two-wheelers	20	3172	292
Passenger cars	32	3172	431
Light-duty Trucks	15	3172	455
Passenger Busses	6	3172	937
Heavy-duty vehicles	27	3172	1197

PROJECT BENCHMARKS: We assume that a number of similar road improvements have been undertaken in Brazil over the last five years and that the expected percentage reduction in fuel use has been determined for each of these activities. However, no information is available on the number, type and emission rates of the vehicles traveling on these different roads.

Road Improvement Project	Improvement in Fuel Use (%)
Road 1	3.4
Road 2	11
Road 3	7.5
Road 4	6.5

PROJECT ANALYSIS TABLE: Project Number TS9

	U.S. Proposal	EU Proposal	Full Technology Matrix	Hybrid Technology Matrix
Does project qualify?	The U.S. proposal does not provide guidance on the transportation sector; therefore, it cannot be determined whether or not the project would qualify. However, if a percentile test distribution based on energy consumption improvements was developed (in this case percent reduction in fuel use per vehicle mile traveled), then project qualification could be determined. For example, if "X" < 25th percentile, then the project would not qualify.	The positive list allows for improvements in transport energy consumption. The project will result in fuel consumption improvements and will qualify.	The project does not involve the use of advanced, non-commercial technologies; therefore, the project would not qualify under the technology matrix. Project developers would have the opportunity to qualify the project under the project-specific approach and it would likely qualify because credit incentives convinced project developers to proceed with the project.	The project does not involve the use of advanced, non-commercial technologies; therefore, the project would not qualify under the technology matrix. Project developers would have the opportunity to qualify the project under the project-specific approach and it would likely qualify because credit incentives convinced project developers to proceed with the project.
Is the project correctly identified as either a free rider or an additional project?	Indeterminate. U.S. proposal needs to provide guidelines on transportation sector project. However, if percentile test is based on energy consumption, then the project would be incorrectly identified as a free rider.	Yes	Possibly, if the project developers choose to utilize the project specific approach.	Possibly, if the project developers choose to utilize the project specific approach.
Number of credits Awarded	Unknown	Not applicable	Unknown	Unknown
Error in credits Awarded	Unknown	Not applicable	Unknown	Unknown

170

METHODOLOGY ASSESSMENT: This additional project is identified as such only by the EU proposal. The project would not qualify under the technology matrix approach but may qualify under the fall-back project specific approach. Under the U.S. proposal, the project is indeterminate.

Thus far, the EU's positive list is the only approach containing language that deals with infrastructure improvement projects. The U.S. proposal has not yet laid out guidelines for the transportation sector so until it does, the qualification and benchmarking analysis will be incomplete. However, like industrial sector projects, the percentile test analysis for transportation projects could be based on a distribution of energy consumption improvements of recent and comparable activities. In this case, a number of highway infrastructure improvement projects have been implemented in the host country and the percent reductions in fuel use from these projects could be used as a distribution for the percentile test. The average percent reductions in fuel use from these projects could then be used as the benchmark for determining credits. If this criterion were used however, the project would not qualify, as the project's five percent reduction in fuel use would be less than most of these other projects, and the U.S. proposal would fail to recognize it as a truly additional project.

This highlight the problems faced by the U.S. methodology when it is applied to projects that only improve the utilization of a specific technology – but not the efficiency of the technology used. For this type of project, the U.S. should either recommend the use of a fall-back approach (such as the project-specific), or develop guidelines for analyzing projects that improve the utilization of a given technology. These guidelines should be specified for each project type within each sector. For this particular type of project, the additionality or threshold test could be undertaken by comparing the percentage reduction in utilization (i.e. fuel usage per km driven). Estimation of the emission credits awarded could be accomplished by comparing the fuel consumption of the project to the conditions before the implementation of the project.

As for the technology matrix, this road improvement project will not utilize advanced, non-commercial technologies so the project would not qualify under this approach. However, because the project would not be implemented without credit incentives, the project would likely qualify under the project specific approach. Emission reduction calculations could be based on vehicle km traveled per year but this information is currently unavailable and would likely be expensive to obtain potentially jeopardizing project implementation.

PROJECT NUMBER: LU1

COUNTRY: Mexico

SECTOR: Land Use-Change and Forestry

PROJECT TITLE: Forest Protection and Management in Six Mexican Communities

PROJECT DESCRIPTION: In the early 1980s, several organizations were formed to share the costs of managing forests and other natural resources between six communities in rural southern Mexico. The organizations are recognized under Mexican law, and since their inception, the community-based organizations have possessed the right to retain profit from the natural resources in their combined community area. Forest activities currently implemented by the organizations include control and designation of community areas, including permanent forest and agricultural areas. The organizations now propose to conduct a project to improve various aspects of forest protection and management activities in the area.

The project area encompasses approximately 50,000 hectares (ha) of land, of which approximately 32,000 ha consist of closed forest. The remaining land encompasses a mix of open forest, agroforestry, permanent and shifting agriculture, degraded or grazed land, restored forest, and tree plantations. The proposed project activities include the rehabilitation of degraded forest through agroforestry and forest plantation establishment, and the prevention of standing forest degradation. Project benefits include conservation of existing carbon stocks and increased carbon sequestration on forest and agricultural land. The project is also expected to result in wildlife and watershed protection. The benefits of the project have been estimated for a 30-year project lifetime.

PROJECT ADDITIONALITY: The project is additional. While no land loss has been suffered yet, rising human populations within the area are expected to increase resource pressures on this land. Without additional financing and support, the community organizations will be unable to continue sustainable forest management and protection practices. Additional financing will allow the community organization to designate the land as protected forest and agricultural land, prevent the use of land for logging, and protect it from the adverse effects of encroaching human populations. This project seeks to obtain this necessary support through an international carbon offset program, and will not occur without such funding.

IMPACTS OF THE PROJECT: Through improved forest management techniques, the project is expected to reduce the loss of forested area and increase biomass on each hectare. Further, the project is expected to increase the importance of forest management within the community, which is currently more focused on agricultural productivity. The project is expected to increase the amount of closed forestland, tree plantations, and land used for agroforestry. It is expected to decrease the amount of open forest, shifting agriculture and degraded or grazed land.

172

The World Resources Institute's Land-Use and Carbon Sequestration (LUCS) model was used to estimate potential carbon sequestration values resulting from the project. Under the project, over 30 years, carbon sequestration will total 5,606,000 t C; that is, the total of 5,434,000 t C in biomass and 172,000 t C in wood products. This is equivalent to 20,555,332 t CO_2; that is, the total of 19,924,666 t CO_2 in biomass and 630,666 t CO_2 in wood products (using a conversion ratio of 44 t CO_2/12 t C).

The GHG estimates for the 50,000 ha of land in absence of the project activities were also projected using the LUCS model over a 30 year period. Input data for the GHG estimates were developed for the project site, based to some extent on measurement results from on-site studies. In absence of the project, it is estimated that sequestration of 4,499,000 tonnes (t) of carbon; that is, the total of 4,424,000 t C in biomass and 75,000 t C in wood products will occur. This is equivalent to 16,496,333 t CO_2; that is, the total of 16,221,333 t CO_2 in biomass and 275,000 t CO_2 in wood products--using a conversion ratio of 44 t CO_2/12 t C. In absence of the project, the amount of closed and open forest will decrease and the amount of permanent and shifting agriculture, as well as degraded or grazed land are expected to increase.[30]

PROJECT BENCHMARK DATA: Data on similar projects in Mexico are unavailable. Projects such as this, in which mixed use land is being managed in order to rehabilitate degraded forest, establish forest plantations, and prevent standing forest degradation, are currently not taking place in comparable forest areas in the region. Instead, data on the rate of carbon sequestration was collected from ten land areas of a similar size (or range) and land type/use mix within the same region. These areas were picked randomly regardless of the type of land use/forestry activities taking place in that area. Estimates of carbon sequestration over the next 30 years (assuming lands are not subject to the forest management practices described for the project and are not subject to increasing demand pressures) were made using the LUCS model.

---

[30] Even in absence of forest management practices, and without suffering a decrease in land area, the 50,000 ha of land in absence of the project will nonetheless sequester carbon, although it will be a lesser volume than that sequestered under the project. It is important to mention that there exists a natural carbon flux between biomass, soils, and the atmosphere. Carbon is taken up not only by trees, but also by understory, litter, soils, and even wood in landfills, for example. Further, agricultural soils can sequester carbon. Shallow plowing and leaving crop residues to decay on the land can replace depleted soil carbon. Further, abandoned croplands, even if they revert to grasslands, will also sequester some amount of carbon in the soil, as long as the lands do not erode. Thus, while the land under the project, which will be more effectively managed, will sequester a greater amount of carbon, carbon sequestration (though a lesser amount) will still occur on the land in absence of the project.

Land/ Forest Area	Total Land Area In Hectares (ha)	Current CO_2 Sequestration in Biomass	Current CO_2 Sequestration in Wood Products	Current Total CO_2 Sequestration (tons)	Total CO_2 Sequestration per ha
1	49,100	15,231,543	260,850	15,492,393	315.5
2	48,000	15,654,984	200,500	15,855,484	330.3
3	50,200	17,987,542	285,645	18,273,187	364.0
4	55,400	17,000,030	274,896	17,274,926	311.8
5	47,200	13,022,354	254,984	13,277,338	281.3
6	43,000	12,365,456	230,542	12,595,998	292.9
7	53,050	16,200,366	295,005	16,495,371	310.9
8	52,300	16,548,665	270,820	16,819,485	321.6
9	44,650	14,998,000	199,250	15,197,250	340.4
10	50,500	16,231,365	250,325	16,481,690	326.4
Average*	493,400			15,776,312 (15,869,526)	319.5

*Weighted averages shown in parentheses.

PROJECT ANALYSIS TABLE: Project Number LU1

	U.S. Proposal	EU Proposal	Full Technology Matrix	Hybrid Technology Matrix
Does project qualify?	The eligibility threshold for land use projects "may require demonstrating divergence from a regional trend." Because there are no similar land use/forestry activities in the region, the carbon sequestration rate of the project is compared to that of a selection of similarly sized land plots in the same region. If X is set at 10th percentile for total CO_2 sequestered, project will qualify (total CO_2 sequestration > 18,273,187). If X is set at 10th percentile for CO_2 sequestered/ha, project will still qualify (CO_2 sequestered/ha > 364.0).	EU has not accepted the inclusion of land use projects under a carbon offset program. Thus, the project does not qualify as additional.	Technology matrix does not include criteria for determining additionality of land use projects. If the market penetration and economic feasibility tests are used, project would not qualify as additional, as it does not involve an advanced, non-commercial technology. This approach does not specify whether the project-specific approach applies for land use sector projects.	Technology matrix does not include criteria for determining additionality of land use projects; however, this project would probably not qualify as additional, as it does not involve an advanced, non commercial technology. This approach does not specify whether the project-specific approach applies for land use sector projects.
Is the project correctly identified as either a free rider or an additional project?	Yes.	No. Land use activities are not included under the EU's positive list.	No. Technology matrix does not address additionality criteria for land use sector projects, and project does not involve an advanced, non-commercial technology. Unclear whether project specific approach may be applied.	No. Additionality criteria for land use projects not given, and project does not involve an advanced, non-commercial technology. Further, it is unclear whether the project specific approach may be applied.
Number of credits Awarded	Under the U.S. approach, baselines for land use projects are based on the current situation. Thus, the credits awarded to this project would be determined by subtracting the total amount of CO_2 sequestered in absence of the project from the total CO_2 sequestered under the project. Thus, credits = 20,555,332 - 16,496,333 = 4,058,999 t CO_2.	Not applicable.	Project does not qualify for credits.	Project does not qualify for credits.
Error in credits Awarded	Unknown.	Not applicable.	Project is awarded no credits, but needs credits to be economically feasible. Project activities will not occur.	Project is awarded no credits, but needs credits to be economically feasible. Project activities will not occur.

METHODOLOGY ASSESSMENT: Many countries involved in the international climate change negotiations voiced a desire to allow industrialized countries to meet part of their emissions targets through land use and forestry projects. The inclusion of this project sector under an international climate change agreement continues to be an area of intense debate between the U.S. and the EU. The EU does not support the inclusion of land use sector projects because the rules and modalities for accounting for this type of project have yet to be fully established. There remain substantial scientific uncertainties and risks surrounding the verification and accountability of carbon sequestration. Further, from the EU's perspective, inclusion of "carbon sinks" would allow developed countries to emit CO_2 in the atmosphere, while large areas in undeveloped countries will be used as a deposit for "carbon garbage."[31] Thus, while we address this land use sector example, it has not been agreed upon whether such projects will be accepted under an international climate change agreement.

This additional land use sector project fails to qualify as such under all approaches except for the U.S. approach. However, guidance for determining additionality for land use sector projects under the U.S. approach is minimal at best. Under the U.S. approach, it is stated that for land use projects involving carbon sinks, the "eligibility threshold would represent activities that are better than the prevailing conditions within a country or region." Further, it states that "since natural variability may cause sequestration areas to vary immensely, the threshold of performance may require demonstrating divergence from a regional trend."[32] However, it is unclear what is meant by "activities", "prevailing conditions", and a "regional trend." In particular it is unclear whether a comparison of the project and the threshold should be made based on a threshold derived from similar activity types or similar land use/forestry plots. Moreover, if similar forest activities are not being conducted in the area, it is uncertain whether it is necessary to find data on similar activities from another country in order to make the comparison or if it is sufficient to compare the project activities to current sequestration on other land areas (even if these areas are not undergoing similar forest management activities) within the region.

In this case study, for example, a comparison could be made to similar forest activities or to any or all activities in the region, whether they are similar to the proposed project activities or not. The project description indicates that no other forest management and protection projects, such as this one, are being conducted in comparable forest areas in the region. Thus, we assume that the "regional trend" is that forest areas would be left "as is," and would be subject to or threatened by increasing populations and resource pressures that lead to actual decreases in carbon stocks. Therefore, in this case, the project data was compared to sequestration data from ten comparable forest areas in the region. However, clearly specified directions for when to use similar activity types

---

[31] SAIC, "*Political Analysis of the U.S. and EU Market Mechanism Proposals: Subtask 1 Final Working Paper,*" December 2000, pgs. 11-12.
[32] SAIC, " *Political Analysis of the U.S. and EU Market Mechanism Proposals: Subtask 1 Final Working Paper,*" December 2000, pg. 11.

instead of similar land plots should be developed, including default procedures for which activities to include if there are no similar activities to use for the comparison.

In addition, the U.S. approach does not provide guidance on the necessary data that should be used for developing an eligibility threshold. For the industrial sector, for example, the U.S. approach clearly indicates that the eligibility threshold should be set at "X" percentile of efficiencies or emissions rates. There is no such comparable language for land use sector projects. In this case study, we thus conducted two separate eligibility threshold tests based on the data for the ten land areas in the region: (1) a comparison to total CO_2 sequestration in tons, and (2) a comparison to total CO_2 sequestered per hectare. In the former case, total CO_2 sequestration resulting from the project is greater than the total CO_2 sequestration at the 10th percentile of the carbon sequestration occurring in the comparable forest areas and the project qualifies. In the latter case, because CO_2 sequestration per hectare under the project is greater than the CO_2 sequestered per hectare at the 10th percentile, the project proves again to qualify as additional. In this case study, it is preferable to use the later approach, which compares the project in terms of CO_2 sequestered per hectare, because the former approach tends to favor larger land areas, where sequestration would naturally be greater. Nonetheless, it is important for the U.S. to clarify an approach for the additionality determination for land use sector projects, including an explanation of the data to utilize for the eligibility threshold.

The process of estimating credits under the U.S. approach is less problematic than undertaking the additionality test. Under the U.S. approach, "baselines are based on the current situation" when calculating credits.[33] Thus, in this case, the total amount of CO_2 sequestered in absence of the project is subtracted from the total amount of CO_2 to be sequestered with the project (4,058,999 t CO_2).

While the U.S. approach successfully identifies this project as additional, there exist many unique characteristics, such as "leakage" and "permanence," which are inherent to land use sector projects and may undermine the accuracy of additionality and baseline determinations. Leakage is defined as "the unexpected loss of anticipated carbon benefits resulting from additional effects of the project's activities outside the project boundaries." For example, a project designed to prevent deforestation may result in persons moving elsewhere and deforesting other land, resulting in little to no additional carbon savings (i.e., activity shifting). Further, a project designed to reduce forest harvesting may result in the increase of forest harvesting in another region to satisfy demand (i.e., market effects). Leakage is typically associated with a loss in carbon, but in some instances, leakage can be positive when projects lead to more carbon benefits than initially estimated. Permanence is defined as "the possibility of a reversal of carbon benefits from either natural disturbances such as fires, disease, pests, and unusual weather events; or from the lack of reliable guarantees that the original land use activities will not return." For this reason, land use projects should not be considered a permanent solution, but

---

[33] SAIC, " *Political Analysis of the U.S. and EU Market Mechanism Proposals: Subtask 1 Final Working Paper*," December 2000, pg. 11.

rather an opportunity to postpone emissions while simultaneously developing policies and measures and other solutions.[34] Thus, there are many uncertainties surrounding land use sector projects. These uncertainties make it difficult to ensure accuracy in additionality and baseline determinations and to standardize methodology. Although a project may be correctly identified as additional, the benefits of the project may, for example, be reversed by individuals increasing harvesting or deforesting elsewhere or lost by an unforeseen natural disturbance, such as fire. Therefore, a positive additionality determination or the award of credits may later prove erroneous if the estimated carbon sequestration is actually reversed through the problem of permanence or lost through the problem of leakage. Aside from the issues of permanence and leakage, the standardization of methodology for land use sector projects is difficult simply because forest carbon stocks are incredibly varied, depending on latitude, climate, ecosystem (i.e. tropical, temperate, boreal), species mix, and soil regime.[35]

A lack of guidance exists under the EU and technology matrix approaches for determining additionality and awarding credits for projects falling within the land use sector. The EU has clearly stated it does not support the inclusion of land use activities under international market mechanisms for the reasons previously discussed. Thus, at this point, no potential project within the land use sector will qualify as additional under the EU approach for the simple reason that the EU has omitted the inclusion of the land use sector from an international carbon offset program or flexible market mechanism approach.

As it currently stands, the technology matrix approaches do not provide any mention of specific additionality criteria for land use projects. Most land use and forestry projects do not include advanced technologies. Thus, this type of project has not been included in previous work on developing the technology matrix. However, the technology matrix should be updated to specify that land use and forestry projects involving advanced technologies or processes may use the technology matrix approach for the evaluation of additionality, and the project-specific approach for the estimation of credits. Finally, it should specify that all other projects that are not applicable to the technology matrix market penetration and economic feasibility tests should use the project-specific approach.

---

[34] Brown, Sandra. "*Land-Use and Forestry Carbon Offset Projects*," Winrock International, July 1999, p. 9.

[35] WRI "*Getting It Right: Emerging Markets for Storing Carbon in Forests*," 1999.

PROJECT NUMBER: LU2

COUNTRY: Russian Federation

SECTOR: Land-use Change and Forestry

PROJECT TITLE: Afforestation of Marginal Agricultural Land in Russia

PROJECT DESCRIPTION: One particular goal of the Russian government is to enhance environmental quality in the region. The Russian Federation has identified an area of land, southeast of Moscow, totaling 450 hectares (ha) on which to conduct an afforestation project. This land consisted of marginal agricultural land. The Russian Federation plans to attain support from a U.S. investor to afforest the entire tract of land in order to manage it as a carbon sink. In addition to afforestation of the land, this project entails the regular application of fertilizer to the afforested land with the end goal of faster growing trees, and a greater average amount of carbon stored per hectare per year. It is predicted that greenhouse gas benefits will result due to the avoidance of carbon dioxide emissions and via carbon sequestration. In addition to sequestering carbon dioxide, it is expected that the project will also result in reduced soil erosion, improved soil nutrient content, and enhanced habitat area for vertebrate and insect species. Under the project, 450 ha will be afforested with broadleaf (i.e., green ash, box elder, and elm) seedlings.

PROJECT ADDITIONALITY: The project is additional. The project will be undertaken specifically with the objective of avoiding carbon dioxide emissions through avoided soil erosion and biomass decay and the sequestration of carbon through tree growth and the uptake of carbon in the soil. Further, the use and application of fertilizer is a new exercise in the area's forest management practices. The Russian Federation could not have considered conducting the land use project without the additional investment obtained from an outside party.

GREENHOUSE GAS IMPACTS OF THE PROJECT: The site is currently composed of marginal agricultural land; this land has been used as rangeland and was never forested. In absence of the project, this land use is projected to continue, with related CO_2 emissions resulting from soil erosion. The project developers assume a constant rate of soil carbon loss of 0.10 t C/ha-yr over the 40-year lifetime of the project. They also assume that there will be no reforestation or land-use change at the project area. Therefore, total annual reference case emissions for the 450 ha of land are estimated to be 45 t C (=0.10 t C/ha-yr * 450 ha). Thus, for a 40 year project lifetime, the total emissions for the 450 ha without the project are estimated to be 1,800 t C. This is equivalent to 6,600 t CO_2 (using the conversion ratio of 44 t CO_2/12 t C).

Under project conditions, it is assumed that afforestation will prevent emissions from soil erosion and result in carbon sequestration in biomass soils. Carbon sequestration values are derived from annual estimates of stemwood biomass growth, expansion factors to account for total phytomass and litter, and annual soil carbon accumulation rates. The project developers estimate that under project conditions, the afforested 450 ha of land will sequester 668 t C annually. Thus, over a 40-year project lifetime, the afforested 450 ha of land will sequester 26,720 t C. This is equivalent to 97,973 t CO2 (using the conversion ratio of 44 t CO2/12 t C).

PROJECT BENCHMARKS: Other similar sized marginal agricultural lands in the region have been afforested with similar tree species. These areas were not afforested in order to sequester carbon or to avoid carbon emissions, but instead to create plantations for the eventual harvesting of commercial timber. The current annual carbon sequestration for these areas are shown below:

Afforested Land	Area (ha)	Annual Carbon Sequestration (t C)	Tons of Carbon Sequestered per ha
#1	500	788	1.6
#2	400	690	1.7
#3	425	672	1.6
#4	530	820	1.5
Total/Average*	1,855	743 (749)	1.6

*Weighted average is shown in parentheses.

PROJECT ANALYSIS TABLE: Project Number LU2

	U.S. Proposal	EU Proposal	Full Technology Matrix	Hybrid Technology Matrix
Does project qualify?	Under the U.S. approach, the eligibility threshold for land use projects "would represent activities that are better than the prevailing conditions within a country or region." If X is set at the 25th percentile for annual carbon sequestration from the data given for similar afforestation activities in the region, then the project will not qualify (X < 820 t C). If X is set at the 25th percentile for t C sequestered per hectare, the project will not qualify (X < 1.7)	EU has not accepted the inclusion of land use projects under a carbon offset program. Thus, the project does not qualify as additional.	Technology matrix does not include criteria for determining additionality of land use projects. If the market penetration and economic feasibility tests are used, this project may qualify as additional. While it does not involve an advanced, non commercial "technology," it does involve an advanced, non-commercial "process" (fertilizer). This approach does not specify whether the project-specific approach applies for land use sector projects.	Technology matrix does not include criteria for determining additionality of land use projects; however, this project would probably not qualify as additional, as it does not involve an advanced, non commercial technology. This approach does not specify whether the project-specific approach applies for land use sector projects.
Is the project correctly identified as either a free rider or an additional project?	No. However, according to the U.S. approach, due to the natural variability of sequestration, the threshold "may require demonstrating divergence from a regional trend," in which case more emissions data on similar marginal agricultural lands in the region is necessary.	No. Land use activities are not included under the EU's positive list.	Indeterminate. If the technology matrix grants additionality to advanced, non-commercial "processes" (fertilizer), then the project may qualify. Additionality criteria for land use sector projects are not currently addressed under the technology matrix.	Indeterminate. If technology matrix grants additionality to advanced, non-commercial "processes" (fertilizer), then the project may qualify. Additionality criteria for land use sector projects are not currently addressed under the technology matrix.
Number of credits Awarded	Project does not qualify for credits.	Not applicable.	Not applicable.	Not applicable.
Error in credits Awarded	Project is awarded no credits, but needs incentive of credits to be economically feasible; thus, afforestation will not occur.	Not applicable.	Not applicable.	Not applicable.

METHODOLOGY ASSESSMENT: This additional land use sector project fails to qualify as such under all approaches. As discussed in the previous land use sector example, guidance for determining additionality and awarding credits to land use sector projects under the U.S. approach is minimal at best. As previously discussed, the U.S. approach states that for land use projects involving carbon sinks, the eligibility threshold should "represent activities that are better than the prevailing conditions within a country or region." Further, the U.S. approach recognizes that the natural variability of land use projects "may cause sequestration areas to vary immensely;" thus, the threshold may then involve "demonstrating divergence from a regional trend."[36] Again, it is unclear, under the U.S. approach, what is meant by a "regional trend." Further, there is a general lack of guidance under the U.S. approach as to how to effectively measure or determine what constitutes "divergence." It is unclear as to whether the project should be compared to similar activities in the region (such as those given) even though those types of activities may be quite rare, or to all land use and forestry activities in the region occurring on similar land areas.

In this particular instance, sequestration data for four other marginal agriculture land areas are given. These lands are similar in area to the project scenario, and have been afforested with similar tree species. It is assumed that the only difference between the project area and the reference scenario areas is the use or disuse of fertilizer. Because the U.S. approach does not provide guidance on the data necessary for establishing an eligibility threshold for land use sector projects (see LU1), two separate eligibility threshold tests were conducted using the data given. Both a comparison to total annual carbon sequestration in tons and a comparison to tons of carbon sequestered per hectare were conducted. In the former case, the project did not qualify because "X" (i.e., 668 t C/yr) was substantially less than the value for the 25th percentile (i.e., 820 t C/yr) of total tons of carbon sequestered annually. Likewise, in the latter case, the project failed to qualify because "X" (i.e., 1.5 t C/ha) was less than the value for the 25th percentile (i.e., 1.7 t C/ha) of data for tons of carbon sequestered per hectare. As in the last case study, the latter approach is deemed a better comparison than the former approach, because the former approach tends to favor larger land areas, where sequestration would naturally be greater. Again, it is important for the U.S. to clarify an approach for the additionality determination for land use sector projects, including an explanation of the data to utilize for the eligibility threshold. Because this project does not qualify under the U.S. additionality test, it will not be awarded credits.

This example illustrates that carbon sequestration data can indeed vary greatly within a country or region. The comparison here, to lands of similar area and afforested with similar tree species, yielded a comparison to lands which, in many cases, sequestered much higher amounts of carbon than the project scenario. It has been argued that accounting for changes in carbon stocks in land use projects is inherently more difficult

---

[36] SAIC, " *Political Analysis of the U.S. and EU Market Mechanism Proposals: Subtask 1 Final Working Paper*," December 2000, pg. 11.

than accounting for carbon emissions in the power sector. Two significant problems are resolution (recognizing small changes in large numbers) and maintaining the infrastructure needed for regular measurement of changes in carbon stocks. Temporal and spatial variability cause high variability in soil carbon estimates at all scales.[37] Further, as noted in LU1, other issues are inherent to land use sector projects that may undermine the accuracy of additionality and baseline determinations; most notably, "leakage" and "permanence." For example, with regards to "permanence," factors such as drought, frost, weeds, foraging animals, insects, infestation, wind and water erosion, fire, and other unanticipated anthropogenic disturbances could damage the afforested land and cause carbon sequestration to be lost or reversed in future years. As stated, these uncertainties make it difficult to ensure accuracy in additionality and baseline determinations and to standardize methodology.

As demonstrated in the previous case study, a lack of guidance exists under the EU and technology matrix approaches for determining additionality and awarding credits for projects falling within the land use sector. If the technology matrix intends for advanced, non-commercial "processes," as well as technologies to qualify as additional, then this project would likely qualify as additional, given that the new fertilizer constitutes an advanced process. Thus, the treatment of advanced processes under the technology matrix approach should be clarified before it can be properly determined whether a land use/forestry sector project qualifies for credit under the technology matrix. In this instance, the additionality determination is indeterminate.

In addition, the procedures for estimating emission reduction credits under the technology matrix should be developed. The U.S. methodology recommends using the project-specific approach. A similar recommendation should be made for baseline development under the technology matrix.

---

[37] Pew Center on Global Climate Change, "*Land Use and Global Climate Change: Forests, Land Management, and the Kyoto Protocol,*" June 2000, p. 11.

PROJECT NUMBER: RS1

COUNTRY: South Africa

SECTOR: Residential

PROJECT TITLE: Construction of Energy-Efficient Homes in South Africa

PROJECT DESCRIPTION: A U.S.-based construction company proposes to join together with a South African company to construct 1,800 new energy-efficient homes in one South African community. These new homes will be built instead of the standard low-cost homes currently built and subsidized by the government in order to address the acute lack of housing in urban communities. By decreasing the use of kerosene and electricity, the project is also expected to reduce local air pollution, improve indoor air quality, and to contribute to technology transfer and capacity building. The estimated project life is 50 years, and the carbon benefits are estimated to result from reduced space heating.

PROJECT ADDITIONALITY: Currently, the construction of energy-efficient homes has not occurred elsewhere in South Africa, with the exception of a planned Activities Implemented Jointly (AIJ) project to build 4,000 Eco-homes in Guguletu in the City of Cape Town. The project developers decided to construct such homes to replace the standard low-cost subsidized homes in order to obtain the credits that would be available under an international carbon offset program. Without the prospect of carbon offset program participation, the project developers would have opted for housing units of a more traditional design. Therefore, the project is additional.

PROJECT EMISSIONS: The carbon benefits of this project are expected to result from reduced space heating over a 50-year period. Without the project, it is estimated that over 50 years, each of the 1,800 standard low-cost homes in the community would have emitted 86.57 lbs of CO_2 per year, totaling 77.91 tons of CO_2 per year for all the houses or 3,895.65 tons of CO_2 for the entire life of the project. With the replacement of these homes with an equal number of energy-efficient homes, each of the 1,800 new homes is expected to emit 42.93 lbs of CO_2 per year, totaling 39.7 tons of CO_2 per year for all the houses or 1,985.1 tons of CO_2 per year during the life of the project.

PROJECT BENCHMARKS: With the exception of the planned AIJ Pilot Phase project, the construction of energy-efficient homes is not occurring elsewhere in South Africa. In addition, no information is available on the construction of similar types of energy-efficient housing in other parts of the Southern African region. The AIJ project proposal estimates that the projected savings in GHG emissions from Eco-homes range in the area of 50-70 percent compared to the baseline. The following presents fictional data on the expected CO_2 emissions of the various government subsidized housing projects initiated in South Africa within the past two years.

Project Name	Number of Homes	CO$_2$ Emissions Per Year/House (lbs)
Guguletu AIJ project	6,000	44.11
Project 2	3,000	87.99
Project 3	2,000	93.18
Project 4	5,000	101.78
Project 5	4,200	69.23
Total/ (Average)	18,700	396.29 (79.26)

PROJECT ANALYSIS TABLE: Project Number RS1

	U.S. Proposal	EU Proposal	Full Technology Matrix	Hybrid Technology Matrix
Does project qualify?	If $X > 20^{th}$ percentile, then the threshold would be at least 44.11 lbs of CO_2 per year and the project would qualify (ER= 42.93 lbs of CO_2 per year). If $X < 20^{th}$ percentile then there is not enough data to determine project eligibility under the U.S. proposal.	The EU positive list allows for demand side management improvements in residential energy consumption; therefore, the project would qualify.	The project does not involve the use of an advanced, non-commercial technology; therefore, the project would not qualify. However, the project developers would have the opportunity to qualify the project under the project specific approach.	The project does not involve the use of an advanced, non-commercial technology; therefore, the project would not qualify. However, the project developers would have the opportunity to qualify the project under the project specific approach.
Is the project correctly identified as either a free rider or an additional project?	Yes, if $X > 20^{th}$ percentile and unknown if $X < 20^{th}$ percentile. However, project emissions are already lower than AIJ project emissions, which represents the most efficient homes in South Africa. It is then likely that the project would qualify regardless of X.	Yes	Possibly if the project developers choose to utilize the project-specific approach.	Possibly if the project developers choose to utilize the project-specific approach.
Number of Credits Awarded	U.S. approach does not clarify whether credits for residential projects are awarded by comparing project to sector average, or the emissions of the activity to be replaced. However, if it is treated similarly to new capacity projects in the power generation sector, credits would be awarded by comparing the project emissions with an average of recent construction activities.	Not Applicable	Under the project specific approach, credits are awarded at a rate of 86.57 lbs of CO_2 per house/per year - 42.93 lbs of CO_2 per house/ per year = 43.64 lbs of CO_2 per house/per year.	It is not clear, whether credits are awarded by comparing project to sector average, or the emissions of the activity to be replaced. However, this replacement project is treated similarly to replacement projects in the power generation sector, and credits are awarded by subtracting the project emissions from the emissions of the homes to be replaced (86.57 - 42.93 = 43.64 lbs CO_2 per house/year).
Error in Credits Awarded	Unknown	Not Applicable	Unknown	Unknown

186

METHODOLOGY ASSESSMENT: This additional project is correctly identified as such under the EU proposal and would likely be identified as such under the U.S. proposal. Under the technology matrix approach, the project would incorrectly be deemed a free rider.

Although the U.S. proposal does not offer specifics on the residential sector, we assume that its general approach to additionality and baseline development applies. According to the U.S. approach for other sectors, the additionality test should involve a comparison with a reference scenario consisting of a set of recent and comparable activities. In this case, we interpreted "recent and comparable" as meaning new housing projects built within the last two years. In this case, five residential housing projects represent the reference scenario. With only five data points, the 20th percentile is the lowest eligibility threshold we can calculate, and at this threshold, the project easily qualifies. To establish a lower threshold (e.g. 5th or 10th percentile) more data points would be needed. However, because the AIJ project represents the best and most efficient housing project in South Africa and this project's emissions are already lower than the AIJ project's, it is likely that the project will qualify regardless of what percentile is used. This example highlights potential problems with the U.S. proposal's percentile test and the data needed to support the test. The U.S. proposal does not indicate whether the percentile will be fixed for all projects or if it will change on a project-by-project basis. In this case for example, if X were set at the 10th percentile for all projects, then more data would be needed to establish the threshold. In addition, if we are to remain true to the "comparable" requirement of the proposal, the reference points used for the threshold should probably consist solely of other eco-housing projects. With only one of the five reference scenario projects being an eco-housing project, performing the percentile test and establishing a threshold becomes impossible.

More guidance is also needed under the U.S. approach on the process for estimating credits for residential projects, particularly regarding the treatment of replacement/retrofit versus new capacity projects. As the project description indicates the new energy-efficient homes will be built to meet new demand and will be used instead of building an equal number of traditional low-cost homes. Under the U.S. approach, as it is applied to the power sector, it is recommended that retrofit projects that provide new capacity should compare project emissions with a sector average of recent activities. Using this approach, credits would be awarded to this project by subtracting the project emissions (42.93 lbs of CO_2 per year) from the average emissions of the reference projects (79.26 lbs of CO_2 per year). As a result, credits would be awarded to the project at a rate of 36.33 lbs of CO_2 per year.

However, it should be noted that the use of CO_2 emissions/house might not provide a very accurate picture of average emissions for this type of project. Typically, housing projects will differ due to variations in house size, climate conditions, construction practices, etc. Perhaps a more meaningful means of estimating the project benefits would be to base the comparison on the percent improvement on emissions per house.

As for the technology matrix approach, this project does not involve an advanced, non-commercial technology but rather improved, non-commercial energy efficient construction designs. The project demonstrates that the technology matrix will reject "low-tech" energy efficiency improvement projects that may result in true emission reductions. The technology matrix may need to be re-examined in an attempt to accommodate this type of project. The advantage of the technology matrix is that it offers the project-specific approach as a fall back and the project developers would have the opportunity to qualify the project using this approach. In this case, since the efficient housing would not have been built without CER incentives, the project would likely qualify under the project specific approach.

PROJECT NUMBER: RS2

COUNTRY: Mexico

SECTOR: Residential

PROJECT TITLE: Sale of High-Efficiency Light Bulbs for Homes

PROJECT DESCRIPTION: Demand for electricity in Mexico is growing by more than 5 percent per year, spawning the need to add 14,000 MW of capacity over a 10 year period. The average price for electricity in Mexico is below long run marginal costs. Major subsidies exist among residential consumers with medium to large consumers subsidizing smaller users. The Mexican Government is committed to eliminating these subsidies and aims to raise the average price for electricity to equal long run marginal costs.

As a means of achieving these goals, the project developers wish to replace approximately 1 million ordinary, incandescent light bulbs with compact fluorescent light bulbs (CFLs) in one Mexican city. The CFL light bulbs require less energy than the ordinary light bulbs to produce similar or better quality lighting, resulting in less electricity generation and fewer fossil fuel emissions. The CFLs also last up to thirteen times longer than the ordinary bulbs, reducing the cost of the bulbs. The CFLs will be imported from a U.S.-based manufacturer. Currently, there are no CFLs being used in Mexico.

PROJECT ADDITIONALITY: This project is a free rider. Mexico has a strong environmental policy framework and has completed a national environmental action plan, which emphasizes improved air quality in urban areas. Although the project is being financed in part through various grants, the project is being undertaken for the aforementioned reasons, not due to the incentive of obtaining credits under a carbon offset program. The project is funded by grants provided to encourage reductions in local and regional pollutants. Indeed, the lighting project is part of the local strategy for improving air quality and reducing demand for electricity. Thus, although a project of this type has never occurred in Mexico before, the project would have been undertaken on a smaller scale without the partial grant funding and the potential sale of credits.

PROJECT EMISSIONS: Without the project, it is assumed that the homes in the Mexican city will continue to use ordinary light bulbs. Currently, the city residents estimate their use of the light bulbs as three hours per day. The baseline condition, then, is the use of 1.0 million ordinary (incandescent) light bulbs for 1,095 hours per year. The ordinary incandescent light bulbs are 100 watts each, and burn at a rate of 0.1 kWh. Using this information, along with the emissions factor of 0.715 lbs CO_2/kwh from the plant serving the community, we derive the emissions rate (without the replacement of the incandescent bulbs).

$ER = (1.0 \text{ mil bulbs})(1,095 \text{ hours/year})(0.1 \text{kwh/year})(0.715 \text{ lbs } CO_2/\text{kwh})$

$ER = (78.3 \text{ million lbs } CO_2/\text{year})(5 \times 10^{-4})$

ER = 39,150 tons CO_2/year

Because the new CFL replacement bulbs are 50 watts each and burn at a rate of .05 kWh, the emissions rate of the project would then be:

$ER = (1.0 \text{ mil bulbs})(1,095 \text{ hours/year})(0.05 \text{kwh/year})(0.715 \text{ lbs } CO_2/\text{kwh})$

$ER = (39.1 \text{ million lbs } CO_2/\text{year})(5 \times 10^{-4})$

ER = 19,550 tons CO_2/year

PROJECT BENCHMARKS: No projects of this type have been initiated or conducted in Mexico. Data on similar projects is also lacking for surrounding countries at this time. Therefore, no data are available to support the development of either country-specific or regional benchmarks.

PROJECT ANALYSIS TABLE: Project Number RS2

	U.S. Proposal	EU Proposal	Full Technology Matrix	Hybrid Technology Matrix
Does project qualify?	The data required to establish a threshold is unavailable either for Mexico or surrounding countries. In this case, it may be necessary to use either a continental or even global threshold to establish additionality.	The EU positive list allows for demand side management improvements in residential energy consumption. Therefore, the project would qualify as additional.	The project involves the use of advanced, non-commercial technology; therefore, the project would qualify as additional.	The project involves the use of advanced, non-commercial technology; therefore, the project would qualify as additional.
Is the project correctly identified as either a free rider or an additional project?	Indeterminate. Data required to establish a threshold is unavailable.	No. The project meets the criteria of the positive list, but the project would have been undertaken even without the incentive of obtaining credits.	No. The project is additional under the technology matrix criteria, but the project would have occurred without the incentive of obtaining credits.	No. The project is additional under the technology matrix criteria, but the project would have occurred without the incentive of obtaining credits.
Number of Credits Awarded	Because this is a retrofit project, if the project should qualify, then credits would be determined by subtracting the project ER from the ER of the light bulbs to be replaced. Thus, credits would be awarded at a rate of 39,150 tons CO_2/year - 19,550 tons CO_2/year = 19,600 tons CO_2/year.	Not Applicable.	Under the technology matrix, credits for replacement projects should use the project-specific approach for baseline development. In this case, credits would be awarded at a rate of 39,150 tons CO_2/year - 19,550 tons CO_2/year = 19,600 tons CO_2/year.	Because this is a replacement project credits would be determined by subtracting the project ER from the ER of the light bulbs to be replaced. Thus, credits would be awarded at a rate of 39,150 tons CO_2/year - 19,550 tons CO_2/year = 19,600 tons CO_2/year.
Error in Credits Awarded	Unknown.	Not Applicable.	Project is a free rider; thus, error is equal to 100 percent of the awarded credits (19,600 tons CO_2/year)	Project is a free rider; thus, error is equal to 100 percent of the awarded credits (19,600 tons CO_2/year)

METHODOLOGY ASSESSMENT: This free rider project was incorrectly identified as additional under the EU and the technology matrix approaches. A qualification of the project was indeterminate under the U.S. approach.

Although the U.S. proposal does not offer specifics on the residential sector, we assume that its general approach to additionality and baseline development applies. In this case, the project in question entails the replacement of 1 million ordinary, incandescent light bulbs with an equal number of more-efficient compact fluorescent light bulbs (CFLs). No projects of this type of have yet occurred in Mexico, and we are told that data on similar projects is lacking for surrounding countries at this time. Therefore, project additionality is indeterminate using the U.S. methodology. In this case, it may be necessary to use either a continental or global threshold to establish additionality. However, the such data would likely introduce errors, as the electricity generation mix, rate of light bulb use, the light bulb burn rate, and other relevant factors differ from region to region. For these same reasons, the use of the emissions rate of CO_2 emissions/year does not provide a meaningful comparison. Projects involving greater bulb usage and relying on carbon intensive electricity use will always result in greater absolute emission reductions than projects using less lighting (per day) and cleaner electricity. Thus, a better approach would be to base the comparison on the percent improvement in energy usage/emissions per home.

Under the U.S. approach, guidance is also needed for evaluating replacement versus new capacity projects during the process of estimating potential emission credits. According to the U.S. methodology, credits are awarded for a replacement project in the power sector by subtracting the project emissions from the emissions of the activity to be replaced. If this light bulb replacement project had qualified as additional, credits would have been determined by subtracting the emissions rate of the 1 million replacement bulbs from the emissions rate of the 1 million light bulbs to be replaced (i.e., 19,600 tons CO_2/year).

This replacement project is incorrectly identified as additional under the EU and technology matrix approaches. Although the project meets the individual additionality criteria of the EU (i.e., demand side management improvements in residential energy consumption) and the technology matrix (use of advanced, non-commercial technology) approaches, the two approaches fail to screen out projects already being implemented due to existing laws. A criterion should be added to all three methodologies that effectively screens out projects already implemented due to existing laws and regulations.

PROJECT NUMBER: RS3

COUNTRY: Russian Federation

SECTOR: Residential

PROJECT TITLE: Energy Efficiency of Seven Apartment Buildings

PROJECT DESCRIPTION: This project involves the renovation and insulation (among other activities) of seven buildings consisting of 424 apartments within a small urban community in the Russian Federation. The buildings are owned by four residential cooperatives, and were built in 1964 and 1965. The building roofs are flat and have suffered many leaks, causing considerable energy losses. The buildings have one-pipe heating systems connected with district heating. No regulation possibilities were available for incoming heating water or for local needs within the buildings.

The main components of the project entail: renovation and insulation of the roofs, wall element joints, entrances and substations, weather-stripping of the windows, installation of heat exchangers, expansion tanks and main pipe control valves, balancing of the heating system, exchange of old and leaking pipes, and the chemical cleaning of the house heating system.

The renovation activities are a joint venture between a U.S. company and a company within the Russian Federation. The partners plan to share any credits generated by the project.

PROJECT ADDITIONALITY: These types of activities to enhance energy efficiency in buildings are occurring extensively in the Russian Federation, and in other parts of the world. However, this project is still sub-economic when evaluated without the potential financing provided through emission credits. When the value of the estimated credits was factored in, the project met the partners' economic feasibility requirements. Therefore, the project is additional.

PROJECT EMISSIONS: Before the project, the apartment buildings drew 1,700 MWh of power from a nearby plant per year, resulting in emissions of 459 tons of CO_2/year. The emissions factor for power supplied by the plant has been estimated at 0.27 tons CO_2/MWh. Therefore, taking into account the fact that the project results in an energy savings of 1,100 MWh/year, the project emissions rate can be computed as follows:

ER = (600 MWh/year)(0.27 tons CO_2/MWh)

ER = 162 tons CO_2/year

The percentage emission improvement of the project is: 65 %

PROJECT BENCHMARKS: There is no available data on this exact combination of renovation and insulation activities in residential buildings that can be used to compare the project to either in the Russian Federation or surrounding countries. However, it should be possible to collect information about other combinations of energy efficiency improvement activities in Russia. A number of hypothetical projects are listed below.

Project	Number of Buildings Included in Project	Energy Savings (MWh/year)	Project Emission Rate (tons CO_2/year)	Emissions Improvement (% CO_2/year)
Project 1: Renovation/Insulation of Flat Roofs; Exchange of Old and Leaking Pipes/New Insulation; Installation of Radiator Reduction Valves	5	670	280	64
Project 2: Installation of New Heating System	7	800	350	62
Project 3: Installation of New Heating System; Installation of Radiator Reduction Valves	4	310	310	59
Project 4: Installation of New Heat Exchangers and Hot Water Circulation System	3	300	258	47
Project 5: Chemical Cleaning of Building Heating System; Installation of Radiator Reduction Valves; Renovation/Insulation of Flat Roofs	10	550	420	54
Total/ (Average)	37	2,630 (526)	1,618 (324)	(47)

PROJECT ANALYSIS TABLE: Project Number RS3

	U.S. Proposal	EU Proposal	Full Technology Matrix	Hybrid Technology Matrix
Does project qualify?	If $X > 20^{th}$ percentile, then the project will qualify (% emissions improvement of the project $> 64\%$ threshold). If $X < 20^{th}$ percentile, then there is not enough data to determine project eligibility under the U.S. proposal.	The EU positive list allows for demand side management improvements in residential energy consumption; therefore, the project would qualify.	The project does not involve the use of an advanced, non-commercial technology or process. These types of renovation activities are occurring extensively in the Russian Federation. Thus, the project would not qualify. However, the project developers would have the opportunity to qualify the project under the project specific approach.	The project does not involve the use of an advanced, non-commercial technology or process. These types of renovation activities are occurring extensively in the Russian Federation. Thus, the project would not qualify. However, the project developers would have the opportunity to qualify the project under the project specific approach.
Is the project correctly identified as either a free rider or an additional project?	Yes, if $X > 20^{th}$ percentile.	Yes.	Possibly, if the project developers choose to use the project specific approach.	Possibly, if the project developers choose to use the project specific approach.
Number of Credits Awarded	Because this project involves retrofitting of existing processes, credits would be determined by subtracting the project ER from the ER of the previous situation. In this case, credits would be 459 - 162 tons CO_2/year.	Not Applicable.	Under the project specific approach, credits are awarded at a rate of 459 tons CO_2/year - 162 tons CO_2/year = 297 tons CO_2/year.	Because this project involves retrofitting of existing processes, credits would be determined by subtracting the project ER from the ER of the previous situation. In this case, credits would be 459 - 162 = 297 tons CO_2/year.
Error in Credits Awarded	Project is additional, so the error in awarded credits is 0.	Not Applicable.	Unknown.	Unknown.

METHODOLOGY ASSESSMENT: This additional project is correctly identified as such under the U.S. and EU approaches, as well as under the technology matrix approaches if the project-specific approach is used.

The analysis of residential projects under the U.S. approach raises several important issues. In particular, a problem arises when comparing projects to a set of recent and comparable activities ("reference scenario") in order to determine additionality. This Russian project involves many different components, such as renovation and insulation of roofs, weather-stripping of windows, and exchange of old and leaking pipes. However, due to the variation in project components, it is difficult to compare the project to other projects because they all very in size and utilize different efficiency measures. For example, the five recent and comparable efficiency improvement activities provided for the threshold analysis of the Russian project involve some, but not all, of the project's components, and the number of buildings differ. In addition, it is likely that each of the reference activities utilize electricity from a different energy mix than that of our specific project, making it meaningless to compare absolute emission reductions (tons CO_2/building). Rather, for the purpose of determining additionality, we recommend using a comparison based in the percentage emissions improvement of each project. Using this approach, the Russian project was deemed to be additional when compared to the 20th percentile of the reference scenario data (i.e., the threshold). In other words, the percent emissions improvement of the project (i.e., 65 percent) was greater than the threshold of 64 percent.

The estimation of emissions credits under the U.S. approach also presents a similar problematic situation. Under the U.S. methodology, it is unclear whether credits for a project involving modifications to existing facilities, would be computed using the emission rate prior to the project or by using a benchmark for comparable facilities. In the former case, credits are easily determined by subtracting the emissions rate after the project from the emissions rate prior to the project (i.e., 459 tons of CO_2/year - 162 tons CO_2/year = 297 tons CO_2/year).

Under the EU approach, projects of this nature will always qualify as additional, as the list specifically allows for demand side management improvements in residential energy consumption. As for the technology matrix approach, this project does not involve an advanced, non-commercial technology or process but rather energy efficiency improvement activities that are widely used in the Russian Federation. Hence, it will fail the additionality test of the technology matrix, but may be able to qualify using the project-specific approach. In this case, credits would be awarded at the rate of 297 tons CO_2/year.

PROJECT NUMBER: CS1

COUNTRY: Philippines

SECTOR: Commercial

PROJECT TITLE: Energy Efficiency and Conservation Measures in Commercial Buildings

PROJECT DESCRIPTION: This project involves a mix of energy efficiency measures in two commercial buildings in the city of Manila. The measures include improved water heater insulation, retrofit of HVAC motors with new high efficiency models, and the installation of ultrasonic humidification systems to replace aging infrared systems. Together, these conservation activities will reduce demand for electricity from the nearby coal-fired power plant. Although, the humidification system can be considered non-commercial in the Philippines, the other technologies/processes have already been used for a number of building projects in the country.

The project is undertaken as a joint effort between the local utility supplying electricity to the two buildings and the owner of the buildings. An American investor has expressed interest in financing part of the project in exchange for the potential emission reduction credits.

PROJECT ADDITIONALITY: This project is not a free rider. Although the energy efficiency measures will reduce energy usage and thus building costs over the long run, the building owner would not have undertaken these specific conservation measures without the prospect of raising additional financing through the sale of credits. In particular, the building owner would not have invested in the advanced humidification system, which in terms of up-front costs is considerably more expensive than the standard systems currently used.

PROJECT EMISSIONS: The overall energy savings of the project are 5,530 MWh each year, with the humidification activity accounting for 1,060 MWh of that total. Prior to the project, the two buildings used 12,873 MWh electricity, resulting in emissions of 11,714 tons of CO_2 each year. The emissions factor for power supplied from the generating plant has been estimated at 0.91 tons CO_2 per MWh. Thus, the project's emission reductions can be computed as follows:

ER = (5,530 MWh/year)(0.91 tons CO_2/MWh)

ER = 5,032 tons CO_2/year

The percent emission improvement of the project is: 57%

PROJECT BENCHMARKS: There is no available data on this exact combination of energy efficiency and conservation measures in commercial buildings in Manila or the rest of the country. Thus, a number of hypothetical projects are listed below.

Project	Number of Buildings Included in Project	Energy Savings (MWh/year)	Project Emission Reductions (tons CO_2/year)	Emissions Improvement (% CO_2/year)
Project 1: improved insulation and new HVAC systems	7	8,722	5,233	38
Project 2: Installation of new heating and lighting systems,	7	800	640	52
Project 3: Improved lighting, ventilation and air conditioning	4	6,780	2,373	44
Project 4: Improved HVAC motors, ventilation system, and water heaters	11	9,103	7,282	49
Project 5: Installation of new heat exchangers and cooling system.	6	3,890	3,539	37
Total/Average	35	29,260 (5,852)	19,067 (3,813)	(45)

PROJECT ANALYSIS TABLE: Project Number CS1

	U.S. Proposal	EU Proposal	Full Technology Matrix	Hybrid Technology Matrix
Does project qualify?	If $X > 20^{th}$ percentile, then the project will qualify, because the % emissions improvement of the project (i.e., 57%) is greater than the 52% threshold. If $X < 20^{th}$ percentile, then there is not enough data to determine project eligibility under the U.S. proposal.	The EU positive list allows for demand side management improvements in commercial energy consumption. Therefore, the project would qualify as additional.	While one individual component of the project (the humidification system) is considered advanced, all energy efficiency and conservation measures included in the project are not considered to be advanced and non-commercial. Therefore, a determination of additionality is indeterminate. Evaluators should apply the project-specific approach instead.	While one individual component of the project (the humidification system) is considered advanced, all energy efficiency and conservation measures included in the project are not considered to be advanced and non-commercial. Therefore, a determination of additionality is indeterminate.
Is the project correctly identified as either a free rider or an additional project?	Yes, if $X > 20^{th}$ percentile.	Yes.	Indeterminate. While one project component involves use of an advanced, non-commercial technology, the other components do not.	Indeterminate. While one project component involves use of an advanced, non-commercial technology, the other components do not.
Number of Credits Awarded	Because this is a replacement project involving improvements to existing commercial facilities, credits would be determined by subtracting the project ER from the ER prior to the project. Thus, credits are awarded at a rate of 11,714 tons CO_2/year - 5,032 tons CO_2/year = 6,682 tons CO_2/year.	Not applicable.	Indeterminate due to an indeterminate additionality qualification. However, if the project were deemed additional, credits would be awarded using the project specific approach, because this is a replacement project involving improvements to existing facilities (i.e., 11,714 tons CO_2/year - 5,032 tons CO_2/year = 6,682 tons CO_2/year).	Indeterminate due to an indeterminate additionality qualification. However, as this is a replacement project involving existing capacity, credits would be determined by subtracting the project ER from the ER prior to the project (i.e., 11,714 tons CO_2/year - 5,032 tons CO_2/year = 6,682 tons CO_2/year).
Error in Credits Awarded	Project is additional; thus, error in credits awarded is 0.	Not applicable.	Unknown.	Unknown.

METHODOLOGY ASSESSMENT: This additional project is correctly identified as such under the U.S. and EU approaches. Under the technology matrix approaches, a definite additionality determination could not be made.

Residential and commercial projects typically involve a combination of various energy efficiency and conservation measures. As a result, this type of project often encounters problems when subjected to standardized baseline approaches such as the U.S. methodology and the technology matrix. For example, we found it difficult to compare this particular project in the Philippines to other "recent and comparable activities" as otherwise required by the U.S. approach. It is virtually impossible to identify a set of projects that involve a similar list of efficiency measures (water heater insulation, retrofit of HVAC motors, and installation of humidification systems) and rely on a similar electricity generation mix. Thus, it is difficult to establish an average emissions rate to which the emissions of the project can be compared. The best option would be to rely on a comparison of the percentage improvement in emissions, rather than a comparison of absolute reductions. Applying this approach to the reference data supplied, the project was determined to be additional when compared to the 20th percentile. In other words, the percent emissions improvement of the project (57 percent) was greater than the threshold of 52 percent.

This project encountered even greater problems when subjected to the technology matrix. Under the technology matrix approach, technologies or processes that are determined to be advanced and non-commercial are deemed to be additional. However, the Philippine project involves a number of individual energy efficiency and conservation improvement measures, of which only one is advanced/non-commercial. The technology matrix methodology, however, does not provide guidance on whether the project should be deemed additional based on its inclusion of one advanced, non-commercial technology, or whether it should not be deemed additional because the majority of involved measures constitute more standard, commercial practices. It is also unclear whether technologies used under one larger blanket project can be viewed as separate entities when being examined for additionality under the technology matrix, or whether all components of a project must be viewed as a whole. Thus, in this particular instance, additionality of the project was deemed to be indeterminate. Further guidance is necessary regarding the treatment of this type of multi-component project.

Projects of this nature will always qualify as additional under the EU's positive list. The positive list specifically allows for demand side management improvements in commercial energy consumption.

PROJECT NUMBER: CS2

COUNTRY: Indonesia

SECTOR: Commercial

PROJECT TITLE: Motor Replacement Project in Commercial Office Buildings in Jakarta

PROJECT DESCRIPTION: This project involves the replacement of 1,673 motors for heating, ventilation, and air conditioning (HVAC) systems with premium efficiency motors. The majority of the motors will be added to HVAC fan systems to control the supply and return fans. Additional exhaust fan motors will be used for cooling tower controls. The motor replacements will be undertaken in five large commercial office buildings in the downtown area of Jakarta. It is expected that the new motors will lead to a reduction in electricity use from the nearby utility, which supplies electricity from a mix of coal and natural gas-fired generation.

The project is undertaken as a joint effort between the owner of the office buildings and the American supplier of the premium efficiency motors.

PROJECT ADDITIONALITY: The project is not a free rider. The premium efficiency motors are considerably more expensive than other standard motors and have not yet entered the Indonesian market. The American supplier of the new motors is selling the technology to the building owner in exchange for the potential emissions credits and as a means of introducing the technology to the Indonesian market.

PROJECT EMISSIONS: The project results in a 32 percent decrease in energy use of each motor installed. Before the replacement project, each motor used 2.62 MWh electricity a year. The new premium efficiency motors would use 1.78 MWh each year. With an emission factor of 0.79 mt CO_2/MWh for the electricity supplied from the local utility, the emission rate of the project would look as follows:

ER = (1.78 MWh)(0.79 mt CO_2/MWh)

ER = 1.41 mt CO_2/year

The emission rate of the situation before the project would be:

ER = (2.62 MWh)(0.79 mt CO_2/MWh)

ER = 2.07 mt CO_2/year

201

PROJECT BENCHMARKS: Data on motor replacement projects in office buildings in Indonesia is not readily available. Instead, we have listed a number of hypothetical projects in Indonesia where motors have been installed in buildings. These projects were undertaken within the last two years.

Project	Number of Buildings Included in Project	Number of Motors	Motor Efficiency (MWh/motor)
Project 1	1	295	2.01
Project 2	4	1,042	2.95
Project 3	2	497	2.57
Project 4	3	834	2.68
Project 5	5	1,301	2.45
Total/ (Average)	15	3,969	12.66 (2.53)

PROJECT ANALYSIS TABLE: Project Number CS2

	U.S. Proposal	EU Proposal	Full Technology Matrix	Hybrid Technology Matrix
Does project qualify?	If $X > 20^{th}$ percentile, then the project will qualify (efficiency of the project > 2.01 MWh threshold). If $X < 20^{th}$ percentile, then there is not enough data to determine project eligibility under the U.S. proposal.	The EU positive list allows for demand side management in the commercial sector; therefore, the project would qualify.	The project involves the use of a new technology that has not yet penetrated the Indonesian market. Thus, the project would qualify as additional under the technology matrix.	The project involves the use of a new technology that has not yet penetrated the Indonesian market. Thus, the project would qualify as additional.
Is the project correctly identified as either a free rider or an additional project?	Yes, if $X > 20^{th}$ percentile.	Yes	Yes	Yes
Number of Credits Awarded	Because this project involves the replacement of existing motors, credits would be determined by subtracting the project ER from the ER of the previous situation. In this case, credits would be awarded at a rate of $2.07 - 1.41 = 0.66$ mt CO_2/year per motor.	Not applicable	Under the technology matrix, credits for projects designed to replace existing technologies rather than meet new demand should use the project-specific approach for baseline development. In this case, credits would be awarded at a rate of $2.07 - 1.41 = 0.66$ mt CO_2/year per motor.	Because this project involves the replacement of existing motors, credits would be determined by subtracting the project ER from the ER of the previous situation. In this case, credits would be awarded at a rate of $2.07 - 1.41 = 0.66$ mt CO_2/year per motor.
Error in Credits Awarded	Project is additional, so the error in awarded credits is 0.	Not applicable.	Project is additional, so the error in awarded credits is 0.	Project is additional, so the error in awarded credits is 0.

METHODOLOGY ASSESSMENT: This additional project is correctly identified as such under all three baseline development approaches.

The U.S. approach does not provide guidance on benchmark development for the commercial sector. However, according to the U.S. approach for other sectors, the additionality test should involve a comparison with a reference scenario consisting of a set of recent and comparable activities. In this case, we interpreted "recent and comparable" as meaning new motor replacement projects undertaken within the past two years. By using this data and applying the U.S. approach, the motor replacement project would qualify as additional regardless of the percentile threshold selected. However, if the percentile is set at a point lower than the 20th percentile, additional data points will be required for comparison.

This project involves a replacement project based on the use of more efficient motors. The U.S. guidance on estimating credits from replacement versus new capacity projects is unclear and should be clarified. According to the U.S. methodology, it appears that credits would be awarded for a replacement project in the power sector by subtracting the project emissions from the emissions of the activity to be replaced. Projects that include new activities or provide new generation would use a sector average (or fossil average in the case of the power sector). This project is a retrofit project. Thus, we subtracted the emissions rate of the project from the emissions rate of the motors to be replaced. Accordingly, the U.S. approach would award credits at a rate of 0.66 metric tons of CO_2/year per motor.

As demand side management projects that improve energy consumption in the commercial sector are included on the EU positive list, this project also qualifies as additional under the EU approach. The two technology matrix approaches would also qualify the project. In this case, the technology matrix awards credits at a rate of 0.66 metric tons of CO_2/year per motor.

In sum, this type of single-component demand side management project appears to be handled equally well by all the baseline approaches. The four baseline approaches correctly identify the project as additional and the credits awarded are the same for each methodology. Once the demand side management and energy efficiency projects involve multi-component activities, the standardized baseline procedures encounter significant problems because of the inherent difficulties of establishing a reference case based on similar project components.

APPENDIX B

This following sections provide a detailed description of the official U.S. approach, the EU approach, and the technology matrix.

United States Proposal

Background

In September 2000, the U.S., through the Environmental Protection Agency, issued a formal proposal that offered a methodology for implementing flexible market-based mechanisms. This proposal included a two-step approach to dealing with free ridership and baseline development. In the first step, a project's eligibility for credits is determined through a free rider test. The second baseline development step forms the basis for quantifying the amount of credits to be awarded.

Free Ridership

On the issue of free ridership, the U.S. proposal turned on what could be called the superior performance concept. Under this proposal, the U.S. indicated that free rider tests should:

- Be robust in their ability to screen out "anyway" tons,

- Recognize a reasonable number of certified emission reductions, and

- Be objective and transparent enough to promote participation and investment in project activities that achieve real reductions and support sustainable development.[38]

The proposal's superior performance concept stated that "a more effective approach for gauging additionality of emissions reductions or removals would be to require that the project activity achieve a level of performance with respect to emissions reduction or enhancements of removals that is significantly better than average compared with recently undertaken activities or facilities."[39] In other words, the superior performance free rider test required that a project activity meet an eligibility threshold that is

---

[38] UNFCCC, Mechanisms Pursuant to Articles 6, 12 and 17 of the Kyoto Protocol, Principles, modalities, rules and guidelines for the mechanisms under Articles 6, 12 and 17 of the Kyoto Protocol Paper No. 4 Unites States of America, September 9, 2000, p. 21. Document # FCCC/SB/2000/MISC.4/Add.1.

[39] UNFCCC, Mechanisms Pursuant to Articles 6, 12 and 17 of the Kyoto Protocol, Principles, modalities, rules and guidelines for the mechanisms under Articles 6, 12 and 17 of the Kyoto Protocol Paper No. 4 Unites States of America, September 9, 2000, p. 21. Document # FCCC/SB/2000/MISC.4/Add.1.

significantly better than average compared to a reference scenario. A reference scenario would consist of a set of recent and comparable activities or facilities in a relevant geographic area. Normally the host country would be the relevant geographic area, but a larger or smaller area could be defined to achieve a more representative reference scenario. This general standard would be elaborated and made more specific in the context of specific project categories. As Figure B-1 demonstrates, a certain percentile of the reference scenario would determine the eligibility threshold. As a subset of the reference scenario, the percentile represents the best performing recent and comparable activities or facilities. It is this percentile that the market mechanism project activity must outperform in order to meet the eligibility threshold and qualify as additional. The requirement that the project activity meet the percentile threshold defined the criteria that the project achieve a level of performance that would be "significantly better than average."

Figure B-1. Percentile Test

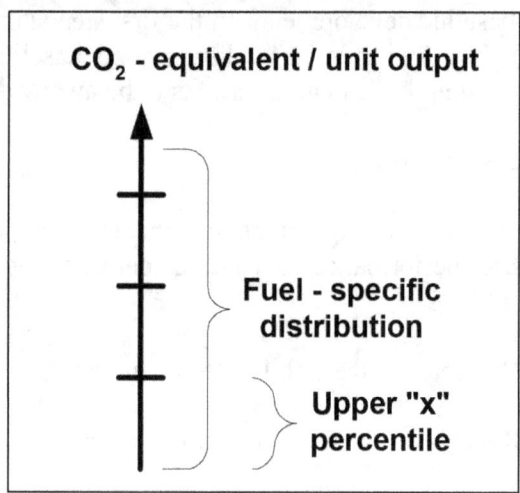

Source: U.S. Environmental Protection Agency
Presentation, SB-13 Lyon, France September, 2000

To illustrate this point and further our understanding of Figure B-1, consider the power sector as an example. Suppose we are developing a new coal plant. To begin the free rider analysis, we would develop our reference case by selecting a set of recent and comparable power plants. In this situation, "comparable" is interpreted as meaning same fuel (i.e. a coal project is compared to other recent coal plants or a gas project is compared to other recent gas plants). Then the emissions rates, expressed as pounds of CO_2/kWh, for each of the coal-fired plants in the reference case would be quantified. Because no two plants are exactly the same, we can expect a range or distribution of emissions rates. We then apply the X percentile to this distribution to quantify the eligibility threshold. For illustrative purposes, we will set X to 10 percent. At the 10th percentile the emission rate might, e.g., correspond to 1.85 lbs CO_2/kWh. This value sets the performance threshold such that our coal project's emission rate would need to be less than 1.85 lbs CO_2/kWh to pass the free rider test.

The process would be the same for a natural gas plant. The reference case would consist of a set of recent and comparable natural gas plants and the emissions rate for each plant would be calculated. The X percentile would then be applied to this distribution to quantify the performance threshold. To illustrate we once again assume X equals 10 percent. In this case, the 10th percentile might correspond to an emissions rate of 0.94 lbs CO_2/kWh. In order for the natural gas project to meet this threshold and qualify as additional, its emissions rate must be below 0.94. However, as we will see shortly, the percentile test would not apply to zero emissions projects, which would include renewables, hydropower and nuclear. These projects would automatically pass the free rider test.

Baseline Development

Baseline development is a key part of the process of quantifying the number of credits a project activity will be awarded. Under this proposal, a project would receive credits after a determination that it met the eligibility threshold and therefore passed the free rider test. Once the eligibility threshold was met, credits would be granted based on what would have happened in the absence of the project activity. "This general standard would be elaborated and made more specific in the context of specific project categories and regions, but should be both realistic and practical. Any proposed baseline that uses a new baseline methodology would need to be approved by the executive board."[40] The proposal argued that the baseline must reasonably represent the emissions or removals that would have occurred in the absence of the project activity and that the reference scenario developed as part of the free rider test would, in many cases, satisfy this criterion. However, for fossil projects, project developers would be allowed to choose from three categories of baselines 1) a fuel-specific average; 2) a fossil average which would include coal, oil, and natural gas plants; and 3) a sector average, which would include fossil, hydropower, nuclear and renewable plants. Under this proposal, the number of credits awarded would be determined by subtracting the project activity emissions from the average emissions of the reference scenario or, in the case of fossil projects, one of the three baseline categories.

Application to Specific Categories

For illustrative purposes, the proposal applied the two-step process to a number of specific project categories, including power generation; industrial practices; methane capture; and land use, land use change and forestry (LULUCF).

Power Generation

The proposal states that

> the eligibility threshold could be met in one of several ways, depending on the reference scenario used, and that methodologies could be developed by looking at the emissions performance of comparable recently undertaken activities or facilities within the Party or region in which the project activity is occurring. For fossil fuel projects, the performance threshold would be measured in terms of emissions per kilowatt-hour generated, and would be set at the [X] percentile of the

---

[40] UNFCCC, Mechanisms Pursuant to Articles 6, 12 and 17 of the Kyoto Protocol, Principles, modalities, rules and guidelines for the mechanisms under Articles 6, 12 and 17 of the Kyoto Protocol Paper No. 4 Unites States of America, September 9, 2000, p. 22. Document # FCCC/SB/2000/MISC.4/Add.1.

recent and comparable activities or facilities and would represent the lowest emissions rates within a reference scenario for each fuel type.[41]

This criterion attempted to account for the fact that every power plant in a combination of recent facilities has a different emissions rate. Rather than simply using the average emissions rate of these recent facilities as the performance threshold, it required that the project activity pass a more stringent performance threshold. To conclude its application of the proposal to the power generation sector, the proposal states that, "[t]he baseline would be set at the weighted-average emissions rate for each fuel type, fossil component, and sector within the reference scenario,"[42] which meant that credits would be awarded based on a comparison of the project activity to an average emissions rate of recent activities.

New Grid-Connected Generation: Figure B-2 demonstrates how the proposal would work in the power generation sector for new grid-connected generation. For a coal project, the figure shows that the project activity's emissions rate would be required to fall below the coal percentile line in order to meet the eligibility threshold and pass the free rider test. To reiterate, the coal percentile would be calculated by applying X percentile to an emission rate distribution of a reference case made up of recent and comparable coal plants. Any coal project falling above the coal percentile line would be considered a free rider and therefore ineligible for credits. For a coal project that passes the free rider test, the project developers would then choose among the three baseline categories – fuel (coal) average, fossil average, or sector average – to calculate its emission credits.

Figure B-2. New Grid-Connected Generation

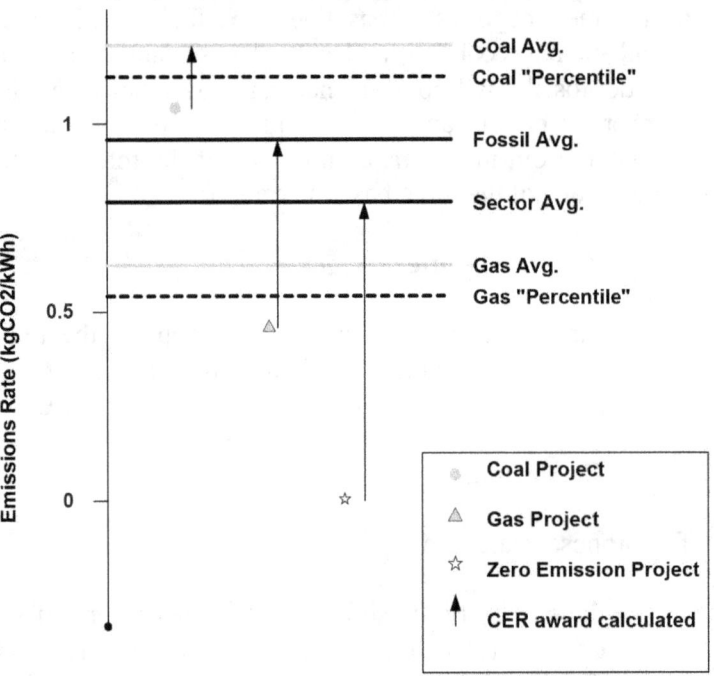

Source: U.S. Environmental Protection Agency Presentation, SB-13 Lyon, France September, 2000

[41] UNFCCC, Mechanisms Pursuant to Articles 6, 12 and 17 of the Kyoto Protocol, Principles, modalities, rules and guidelines for the mechanisms under Articles 6, 12 and 17 of the Kyoto Protocol Paper No. 4 Unites States of America, September 9, 2000, p. 23. Document # FCCC/SB/2000/MISC.4/Add.1.
[42] Ibid.

It can be assumed that project developers will always choose the category that yields the highest emission rate. In the case of a coal project, the fuel (coal) category will have the highest average rate. Hence, the example in Figure B-2 shows that the coal project would use a fuel (coal) average to calculate emission credits. In this example, the reference case from the free rider test would be used to calculate the coal average.

In contrast, a natural gas project would be required to fall below a gas percentile eligibility threshold to pass the free rider test. Like the coal percentile, the gas percentile would be calculated by applying X percentile to an emissions rate distribution from a reference case made up of recent and comparable natural gas plants. The project developers would then choose a baseline from among the three baseline categories [fuel (gas) average, fossil average, or sector average]. In this case, the project developers would presumably choose the fossil category since it would yield the highest average emission rate. Hence, Figure B-2 shows that credits would be awarded by subtracting the gas project's emissions from a fossil average. This fossil average would be a combination of recent coal, oil, and natural gas facilities. Again, if a natural gas project's emissions fall above the gas percentile line, it would not meet the eligibility threshold and would then fail to qualify for credits.

For a zero emissions project, which could include renewables, hydropower, and nuclear, there is no percentile test or eligibility threshold. In the absence of a percentile test, these types of projects automatically pass the free rider test and would be eligible for credits. However, unlike the two previous examples, project developers do not get to choose which baseline category to use. As Figure B-2 demonstrates, a zero emissions project must use the sector average to calculate credits. The sector average would be a combination of recent fossil, renewable, hydropower, and nuclear facilities.

Figure B-3. Coal to Gas Retrofit

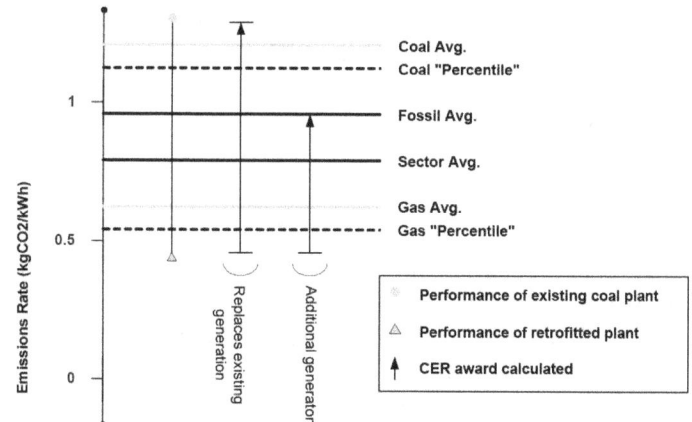

Source: U.S. Environmental Protection Agency Presentation, SB-13 Lyon, France September, 2000

Fuel-Switching: Figure B-3 illustrates how the proposal would work in the power sector for a fuel-switching project. In this example, a coal to gas retrofit project is examined for replacement and new generation. The figure demonstrates that the emissions of such a project would need to fall below a gas percentile line to meet the performance threshold and pass the free rider test. However, the number of credits awarded would differ based on whether the project would replace existing generation or add new generation. As a replacement project, the figure shows that credits would be awarded by subtracting the project emissions from the emissions of the coal plant that the retrofit project would be replacing. Alternatively, if the retrofit project provides new generation, the figure demonstrates that credits would be awarded by comparing the project emissions with a fossil average of recent coal, natural gas, and oil facilities.

New Off-Grid Generation: Figure B-4 illustrates how the proposal would work for new off-grid generation projects. These projects would often be located in remote areas where energy needs are supplied by individual units such as diesel generators or kerosene lamps. In this example, a renewable project is compared to current practices (e.g. kerosene/diesel). As we saw in Figure B-2, renewable, or zero emission, projects do not have a percentile test to pass. They automatically pass the free rider test and in this example, credits would be calculated by comparing the project with the emissions rate of the kerosene or diesel units.

Figure B-4. New Off-Grid Generation

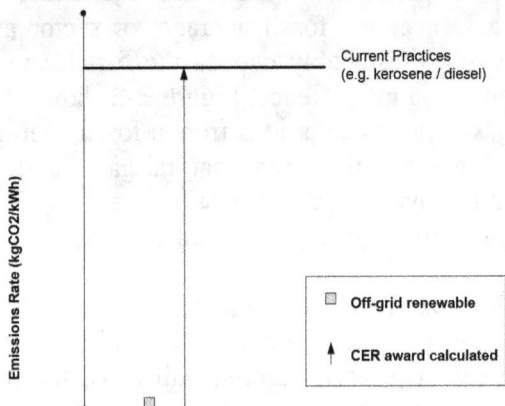

Source: U.S. Environmental Protection Agency Presentation, SB-13 Lyon, France September, 2000

Industrial Practices

The performance of energy-intensive industrial practices, such as production of steel and cement, could be measured in terms of energy use or GHG emissions per output of intermediate products. Like power generation, the eligibility threshold would be set at "[X] percentile of the lowest energy consumption or GHG emissions per unit of output for facilities in the reference scenario, with baselines determined as the weighted-average emission rate per unit of output."[43]

Methane Capture

Methane capture project activities would also be required to meet the "significantly better than average" threshold. For example, if methane capture at landfills was not standard practice – as is the case in many countries – then by definition the threshold would be any capture project that is better than the current situation. The baseline would then be the previously existing condition – i.e., no capture – and the project would calculate credits based on the amount of gas captured.

---

[43] UNFCCC, Mechanisms Pursuant to Articles 6, 12 and 17 of the Kyoto Protocol, Principles, modalities, rules and guidelines for the mechanisms under Articles 6, 12 and 17 of the Kyoto Protocol Paper No. 4 Unites States of America, September 9, 2000, p. 23. Document # FCCC/SB/2000/MISC.4/Add.1.

Land Use, Land Use Change and Forestry (LULUCF)

For sinks, the eligibility threshold would represent activities that are better than the prevailing conditions within a country or region. Since natural variability may cause sequestration areas to vary immensely, the threshold of performance may require demonstrating divergence from a regional trend. For the purpose of calculating credits, baselines would then be based on the current situation.

Flexible Crediting Periods

The proposal also envisioned flexible crediting periods. Project developers, at the time a project activity is registered, would select either a stable baseline over a fixed crediting period or a baseline that is periodically updated over an indefinite crediting period. For a fixed baseline and crediting period project, credits would no longer accrue at the conclusion of the crediting period. For a variable baseline project with an indefinite crediting period, credits would continue to accrue provided that the project continues to meet updated threshold and baseline criteria.

Provisions for Sinks Projects

For projects designed to enhance removals by sinks, the process for selecting a crediting period would be a little different from other GHG emission reduction projects. Project developers would be required to propose a crediting period during which the sequestered carbon would remain sequestered. Project developers would also be required to address the issue of permanence by identifying modalities to account for the possibility that sequestered carbon held in forests or soils might be released into the atmosphere prior to the end of the crediting period. All new methodologies related to crediting periods for sinks would be subject to approval by a market mechanism executive board. For sinks projects, the proposal envisions the executive board as having two key tasks:

1. Approval of carbon accounting methodologies to determine climate benefits of sequestering a ton of carbon for various lengths of time in comparison to climate benefits from a ton of emission reductions from the energy or other sectors. Successful sequestration for longer periods would receive more credits than sequestration for shorter periods.

2. Approval of modalities to ensure that credits from sequestration projects reflect the project's actual climate benefits. Carbon accounting methodologies could address permanence by crediting sequestration fully as it occurs and requiring that premature release of carbon dioxide be made up through liability or insurance measures. Alternatively, credits could be issued incrementally for each year sequestration is maintained.[44]

---

[44] UNFCCC, Mechanisms Pursuant to Articles 6, 12 and 17 of the Kyoto Protocol, Principles, modalities, rules and guidelines for the mechanisms under Articles 6, 12 and

Leakage

The proposal also called on project developers to account for potential significant changes in emissions or removals, reasonably attributable to the project, that are likely to occur outside the project activity. These changes were to be monitored and incorporated into the calculation of credits during certification; however, the proposal did not address how this would be accomplished. In practice it can be very difficult to identify, quantify and even determine ownership of outside project-related leakage.

Monitoring

The final piece of the proposal required project developers to propose a monitoring plan at registration. The plan was to use a methodology previously approved by the executive board or an alternative methodology, which would be subject to executive board approval. The monitoring methodology needed to be sufficiently rigorous to provide an accurate calculation of emissions and/or removals. If developing a rigorous methodology was not feasible, the project developers then must submit a conservative estimate that neither underestimates emissions, nor overestimates removals. The monitoring plan must encompass all emissions or removals occurring within the project boundary and account for leakage using the above criteria as a guide (i.e. emissions or removals that are significant and reasonably attributable to the project activity).

Summary

In short, the U.S. market mechanism proposal was designed to operate as follows. First, a reference case consisting of recent and comparable facilities or activities would be selected and then a set percentile, representing the best performing facilities or activities of this reference case, was to be used to determine a proposed project's free rider status. By using this percentile, the U.S. proposal required the project activity to be "significantly better than the average" to qualify as additional. If the project passed this eligibility threshold, it would be considered additional and would move to the next step of baseline development. The project's baseline was to be determined and credits were to be awarded by calculating the difference between the project's emissions and the average emissions rate of the reference case. For fossil projects, credits were to be awarded by calculating the difference between the project's emissions and a fuel average, a fossil average, or a sector average.

The U.S. argued that this proposal was cost-effective and provided a high level of certainty for project developers, while also maintaining environmental integrity though a free rider test that maintained a significant level of stringency. Moreover, the U.S. argued that this proposal was technology neutral and supported the sustainable development goals of host countries through appropriate technology transfer. The

---

17 of the Kyoto Protocol Paper No. 4 Unites States of America, September 9, 2000, p. 24. Document # FCCC/SB/2000/MISC.4/Add.1.

proposal suggested that this superior performance approach would achieve several objectives simultaneously:

- It served as a reasonable proxy for "investment additionality." Under this approach, it could be reasonably assumed that an investment to achieve a level of superior performance would less likely be undertaken without the promise of emission credits. (A project that achieves only average performance is more likely to have been built anyway.)

- By using an average of *recently* undertaken activities or facilities as a measure for credits, it forced the performance of projects to continually improve. In other words, additional projects of today would eventually fall into the category of recently undertaken activities and be used to calculate the average performance in the future, thereby forcing out older technologies in favor of newer and presumably cleaner technologies. (See Figure B-5)

- It focused on practical, comparable activities allowing for emission reductions across a variety of project categories and differing conditions among countries and regions.

- With an objective performance criterion, the superior performance approach reduced the costs and time for project development and increased the predictability of market mechanisms, creating more certainty for investors and promoting more projects.[45]

Figure B-5. Updating of Performance and Baseline Benchmarks Over Time

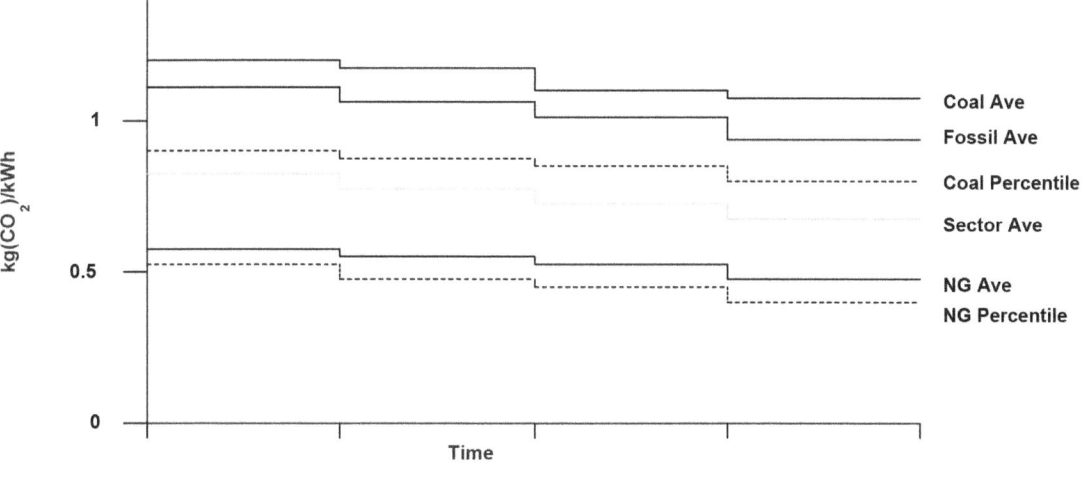

Source: U.S. Environmental Protection Agency Presentation, SB-13 Lyon, France September, 2000

---

[45] UNFCCC, Mechanisms Pursuant to Articles 6, 12 and 17 of the Kyoto Protocol, Principles, modalities, rules and guidelines for the mechanisms under Articles 6, 12 and

Although there was much more detail in this U.S. proposal than any other country's proposal, with the possible exception of the EU, a number of issues still needed to be addressed and defined. For example, the proposal did not define what was meant by "recent" (i.e. three years, five years, ten years, etc.) facilities or activities, or what was meant by "comparable" facilities or activities. In addition, the value of the percentile for the eligibility threshold was not defined. Would it have been a fixed percentile for all projects or would it change from project type to project type or even project to project?

However, despite these questions, it appears that all of NETL's advanced technologies were likely to qualify under this proposal, including clean coal technologies. However, NGOs and environmental groups were critical of the proposal. They argued that the proposal favored continued use of fossil fuels like coal and natural gas over renewables and that nuclear power would have qualified for credits under the proposal. In addition, these groups argued that too many "anyway" reductions or non-additional projects would have received credits as well.

---

17 of the Kyoto Protocol Paper No. 4 Unites States of America, September 9, 2000, p. 22. Document # FCCC/SB/2000/MISC.4/Add.1.

European Union Proposal

Background

The EU feels that the principles, rules, modalities and guidelines for market-based mechanisms should allow maximum environmental and climate benefits to be joined in a cost-effective manner, and provide opportunities for cooperation between countries. Technical work by experts will be crucial in providing a basis upon which to make sound judgments regarding the methodologies and technical processes that will guide the operation of the mechanisms. From the European perspective, a priority for such a task is the development of rules for establishing baselines, against which the effects of project activities may be assessed. Another priority is the development of guidelines for ensuring a transparent and convenient system of tracking and recording transfers under the mechanisms. Further, the augmentation of a comprehensive compliance system is of utmost importance for successful operation of the market mechanisms.[46]

The European anticipation is that international action to meet emission reduction commitments will require cooperation between governments at all levels, industry and other groups in managing the expansion and distribution of environmentally sound technologies in all sectors. Developed countries are proposed to hold responsibility for maintaining a national system for accurate monitoring, verification, and accountability of all emissions covered by the trading scheme, as well as a publicly accessible national registry system recording all pertinent details.[47]

From the European perspective, it is important to encourage the use of renewable energy technologies under the market mechanisms, and help such technologies to compete with fossil fuel technologies. In the long term, the market mechanisms should include a sustainable development process in countries that are still experiencing growth and that require assistance for environmental and social problems. Focused attention is necessary for market mechanism negotiations, or they may produce a program that results in much for developed nations and little for developing countries.[48]

---

[46] United Nations Framework Convention on Climate Change (UNFCCC), Subsidiary Body for Scientific and Technological Advice (SBSTA) and Subsidiary Body for Implementation (SBI), "5th Conference of the Parties on Climate Change, 11th Sessions of SBSTA and SBI: EU Statements," Bonn, November 1999.

[47] United Nations Framework Convention on Climate Change (UNFCCC), Subsidiary Body for Scientific and Technological Advice (SBSTA) and Subsidiary Body for Implementation (SBI), "5th Conference of the Parties on Climate Change, 11th Sessions of SBSTA and SBI: EU Statements," Bonn, November 1999.

[48] Globe EU, "Globe EU Policy Paper Followed by a Report on the Briefing with Jos Delbeke, EU Chief Negotiator on Climate Change," Globe EU Campaign on Climate Change, August 2000, http://USers.swing.be/aude.lemens/cop6.htm

Baseline Development– The EU's "Positive List" of Technologies

To guard against rich countries 'dumping' out-of-date technologies on developing nations, the EU suggests that in the primary phase of the market mechanisms, only projects based on a "positive list" of safe, environmentally sound, clean technology projects should be able to obtain emission credits.[49] The proposed positive list is presented in Table B-1. The first column of this table shows the three main categories of technologies included in the positive list (renewables, energy efficiency, and demand side management), while the second column shows the individual technologies within these three categories.

Under the category for energy efficiency, the EU further specified that fossil power plant projects would be eligible under the market mechanisms only if the following criteria were met:

- New Plants: If the plant has a minimum efficiency of 55 percent for plants larger than 300 MW and 52 percent for plants smaller than 300 MW

- Rehabilitation of Existing Plants: If the project introduces a technology change that leads to an increase in overall plant efficiency of at least 5 percent.

Table B-1. The EU's Proposed Positive List of Technologies

Main Technology Categories	Individual Technologies
Renewables	Solar
	Wind
	Sustainable Biomass
	Geothermal heat and power
	Small-scale hydropower
	Wave and tidal power
	Ambient heat
	Biogas
Energy Efficiency	Advanced technologies for combined heat and power installations and gas fired power plants
	Significant improvements in existing energy production
	Advanced technologies for, and/or significant improvements in industrial processes, buildings, energy transmission, transportation and distribution
	More efficient and less polluting modes of mass and public transport (passenger and goods) and improvement or substitution of existing vehicles
Demand Side Management	Improvements in residential, commercial, transport and industrial energy consumption.

---

[49] European Commission, "Outcome of Climate Change Negotiations in Lyon, France, 4-15 September, 2000 (Press Release)," September 1, 2000, http://europa.eu.int/comm/environment/press/bio00172.htm

216

The EU recommends that each project considered under the market mechanisms undergo an independent Environmental Impact Assessment (EIA), to be commissioned and financed by the project participant. Each EIA should include social impacts, carried out in accordance with existing rules, standards, and legislation of the host country. In absence of such existing rules, the project participant should follow appropriate international guidelines and good practice such as Organization for Economic Cooperation and Development (OECD) guidelines on EIAs.[50]

---

[50] United Nations Framework Convention on Climate Change (UNFCCC), Subsidiary Body for Scientific and Technological Advice (SBSTA) and Subsidiary Body for Implementation (SBI), "Mechanisms Pursuant to Article 6, 12, and 17 of the Kyoto Protocol," (FCCC/SB/2000/MISC.4/Add.2/Rev.1), September 14, 2000.

The Technology Matrix

The technology matrix approach, modified and developed by NETL, consists of a selected list of greenhouse gas abating technologies that correspond to with the sustainable development goals of the host country. The technology matrix represents a cost-effective, transparent, and reasonably accurate approach to quantifying greenhouse gas emission reduction project baselines. It is similar to other benchmarking approaches but with the addition of an effective, rigorous test to eliminate free rider projects. It also specifically addresses the problem of which technologies to include in the benchmark/baseline. The technologies qualify for the list by passing rigorous tests of the candidate technology's economic feasibility and market penetration in the host country. These tests are a means of weeding out business-as-usual or free rider projects. In general, only advanced, non-commercial technologies are likely to pass the test and qualify for inclusion in the matrix.

Economic Feasibility Test

This test involves a comparison of a specific candidate technology's costs to the cost of alternative commercial technologies in a selected country, to determine the candidate technology's commercial viability. In addition to accounting for the cost of implementing the technology itself, factors to be considered in determining a candidate technology's economic feasibility should include energy costs, environmental regulation, tariff structures, etc. Other considerations to be taken into account include whether construction costs can be predicted with reasonable certainty and whether the operational performance of the technology can be guaranteed. If the technology proves unable to compete with current market technologies - in other words, the technology is not commercially viable - it would pass this test and qualify for inclusion in the matrix. Technologies that are likely to pass the economic feasibility test include renewable technologies such as solar and wind, integrated gasification combined cycle technologies, and integrated gasification fuel cell technologies.

Market Penetration Test

Some technologies may prove to be commercially viable but still face certain non-financial barriers to implementation in select countries. These barriers could include risks associated with installing and operating locally unknown technologies, institutional barriers or internal organizational structures that discourage investment in energy sector improvements, or poorly functioning capital markets that prevent new technologies from being adopted. If a commercial technology were shown to have a weak market penetration rate in a certain country, then the technology would still qualify for inclusion in the matrix.

Ideally, the economic feasibility and market penetration tests should work together to qualify technologies for inclusion in the matrix. However, in some instances only one of the tests may be sufficient to qualify a particular technology.

Establishing the Baseline Under the Technology Matrix

Once a technology qualifies, a benchmark is developed for that specific technology based on the emissions performance of a counterfactual technology(ies). The counterfactual technology represents the technology most likely to be utilized, if the corresponding advanced-technology project were to be foregone. There are three basic steps to estimating the benchmark. First, the most likely alternative to the project must be defined in a qualitative manner (i.e. what is the counterfactual technology?). Second, the data required to quantify the benchmark must be collected for each technology/country combination. Finally, the collected data is analyzed, and used to compute the benchmark (i.e. the baseline against which the project emissions will be compared). Once the baseline is established, utilizing the technology matrix is a straightforward process for project developers. To qualify for emission credits, project developers would simply demonstrate that the proposed project technology is included in the matrix. Then the amount of credits to be awarded to the project would be determined by subtracting the project's emission rate from the stipulated benchmark.

Temporal Considerations. As time passes, the economic performance, technological capabilities, and energy intensity of a nation is likely to change. As a result, the list of pre-qualified technologies should be updated regularly, preferably every five years, to capture the impact these changes may have on the individual technologies. If this periodic review reveals that individual technologies are no longer additional, they should be removed from the matrix and added to the activities that make up the baseline. Similarly, the technology baselines also should be updated every five years to account for the introduction of new technologies and other changes that may influence the composition of the benchmark groups used to establish the baselines. An example from the power generation sector of any given country can be used to illustrate this point. An initial group of existing power plants would be selected as best representing the "typical" counterfactuals for projects using a qualifying technology; the average heat rate or emissions rate for this benchmark group would be applied to the first group of projects qualifying under the technology matrix. However, after five years, a new benchmark group, reflecting changes and/or improvements in power plant technology, would be used as the basis for a new benchmark to be applied to all new projects implemented as of year six. In a similar fashion, a new benchmark group of existing power plants would be used to establish a new benchmark at each following five-year interval. Moreover, the baselines developed from the original counterfactual power plants would be updated every five years as well. The development of these plants would be traced over time to account for changes in heat rates and wear and tear on the equipment. In this way, the power plants and technologies originally selected for developing the stipulated baseline will continue to serve as the benchmark throughout the life of the first group of qualifying projects. Table B-2 presents a sample technology matrix for several countries/technology combinations, illustrating the element of time in baseline development for the technology matrix.

219

Table B-2. Sample Technology Matrix with Initial and First Two-Year Baseline Updates

Countries		India			China			Argentina		
Qualifying Technologies		Year 1	Year 6	Year 11	Year 1	Year 6	Year 11	Year 1	Year 6	Year 11
Coal-Fired IGCC	BMG a	B_a	B_{a+5}	B_{a+10}	B_a	B_{a+5}	B_{a+10}	B_a	B_{a+5}	B_{a+10}
	BMG b	---	B_b	B_{b+5}	---	B_b	B_{b+5}	---	B_b	B_{b+5}
	BMG c	---	---	B_c	---	---	B_c	---	---	B_c
Solid Oxide Fuel Cells	BMG a	B_a	B_{a+5}	B_{a+10}	B_a	B_{a+5}	B_{a+10}	B_a	B_{a+5}	B_{a+10}
	BMG b	---	B_b	B_{b+5}	---	B_b	B_{b+5}	---	B_b	B_{b+5}
	BMG c	---	---	B_c	---	---	B_c	---	---	B_c
Phosphoric Acid Fuel Cells	BMG a	B_a	B_{a+5}	B_{a+10}	B_a	B_{a+5}	B_{a+10}	B_a	B_{a+5}	B_{a+10}
	BMG b	---			---	B_b	B_{b+5}	---		
	BMG c	---	---		---	---		---	---	
Molten Carbonate Fuel Cells	BMG a	B_a	B_{a+5}	B_{a+10}	B_a	B_{a+5}	B_{a+10}	B_a	B_{a+5}	B_{a+10}
	BMG b	---	B_b	B_{b+5}	---	B_b	B_{b+5}	---	B_b	B_{b+5}
	BMG c	---	---	B_c	---	---	B_c	---	---	B_c
Proton Exchange Membrane Fuel Cells	BMG a	B_a	B_{a+5}	B_{a+10}	B_a	B_{a+5}	B_{a+10}	B_a	B_{a+5}	B_{a+10}
	BMG b	---	B_b	B_{b+5}	---	B_b	B_{b+5}	---	B_b	B_{b+5}
	BMG c	---	---	B_c	---	---	B_c	---	---	B_c
Photovoltaics	BMG a	B_a	B_{a+5}	B_{a+10}	B_a	B_{a+5}	B_{a+10}	B_a	B_{a+5}	B_{a+10}
	BMG b	---	B_b	B_{b+5}	---	B_b	B_{b+5}	---	B_b	B_{b+5}
	BMG c	---	---	B_c	---	---	B_c	---	---	B_c
Pressurized Fluidized Bed Combustion	BMG a	B_a	B_{a+5}	B_{a+10}	B_a	B_{a+5}	B_{a+10}	B_a	B_{a+5}	B_{a+10}
	BMG b	---	B_b	B_{b+5}	---	B_b	B_{b+5}	---	B_b	B_{b+5}
	BMG c	---	---	B_c	---	---	B_c	---	---	B_c

A Back-Up Methodology

Exclusive reliance on any one baseline methodology could result in lost opportunities; therefore, a flexible approach to baseline development protocols is recommended. If a technology fails to qualify for inclusion in the matrix, the technology matrix is designed to employ the project-specific approach as a back-up methodology. The technology matrix is set up to provide a relatively inexpensive method of qualifying and benchmarking projects that utilize advanced non-commercial technologies. However, the technology matrix, if used exclusively, would automatically *disqualify* all projects that utilize standard commercial technologies. While many such projects are in fact likely to prove to be free riders, there will no doubt be exceptions to this rule. Therefore, a back-up methodology is needed to ensure that the technology matrix would not in and of itself eliminate all commercial technology projects from participation in the market mechanisms. Under the technology matrix approach, project developers would always be afforded the opportunity to use the project-specific approach if they cannot use the technology matrix to qualify their projects.

To ensure appropriate selection between the project-specific and the technology matrix approaches under a flexible protocol concept, several guidelines were proposed. Table 2 summarizes these guidelines, as they would relate to the electricity generation sector. As illustrated in row one of Table B-3, the technology matrix should be the default procedure for analyzing all projects involving the installation of new generating capacity utilizing one of the qualifying technologies. (Table B-3 identifies three exceptions to this rule.) For all projects involving non-qualifying technologies or conventional technologies, the project-specific approach must be utilized. However, projects involving the retrofitting of advanced, qualifying technologies may utilize the technology matrix to determine free ridership while utilizing the project-specific approach to establish the baseline.

221

Table B-3. Criteria for Selecting an Approach to Baseline Development for the Electricity Sector

Project Type	Corresponding Approach	Exceptions
Projects involving the installation of new capacity, and utilizing advanced qualifying technologies	Modified Technology Matrix	1. Projects to be implemented in host countries without qualifying technology lists/sector benchmarks must use the project-specific approach. 2. Project developers may choose to use the project-specific approach to estimate the baseline if they can demonstrate that the result is more accurate. 3. Projects designed to replace existing capacity rather than meet new demand should use the project-specific approach for baseline development if the capacity to be replaced can be readily identified.
All projects utilizing conventional non-qualifying technology.	Project-specific	1. Projects involving the installation of new capacity to meet new demand should use a sectoral benchmark for baseline estimation , unless the project developers choose to use the project-specific approach and can demonstrate that the result is more accurate.
Projects involving the retrofitting of advanced qualifying technologies to existing facilities, with not resulting changed in capacity.	Modified technology matrix free rider test; project-specific to estimate the baseline.	None

www.ingramcontent.com/pod-product-compliance
Lightning Source LLC
Chambersburg PA
CBHW081113170526

45165CB00008B/2431